38. Colloquium der Gesellschaft für Biologische Chemie
9.–11. April 1987 in Mosbach/Baden

Molecular Basis of Viral and Microbial Pathogenesis

Edited by
R. Rott and W. Goebel

With 90 Figures

Springer-Verlag
Berlin Heidelberg New York London Paris Tokyo

Professor Dr. RUDOLF ROTT, Institut für Virologie der Universität, Frankfurter Straße 107, 6300 Gießen, FRG

Professor Dr. WERNER GOEBEL, Institut für Genetik und Mikrobiologie der Universität, Röntgenring 11, 8700 Würzburg, FRG

ISBN-13: 978-3-642-73216-4 e-ISBN-13: 978-3-642-73214-0
DOI: 10.1007/ 978-3-642-73214-0

Softcover reprint of the hardcover 1st edition 1988

2131/3130-543210

Preface

Elucidation of the mechanisms of pathogenesis underlying the diseases caused by viruses and bacteria has fascinated scientists for many years in two ways. Firstly, these pathogenic agents represent relatively simple biological systems for the study of basic biological processes such as replication, gene regulation, genetic variability and host-pathogen interactions. Secondly, process in this field is valuable in a practical sence, since it can help in the control of these diseases. The availability of new genetic and immunological techniques, especially recombinant DNA methods and monoclonal antibody technology, has provided powerful tools for unravelling the genetic, biochemical and immunological basis of viral and microbial pathogenesis. Molecular cloning has allowed the isolation of single genes or groups of genes related to phenotypes which appear to be immunologically important for pathogenesis. The specific elimination of such genes from the complex genomes of the pathogens can now be achieved with similar genetic techniques. These genetic studies have provided additional information on the role played by specific phenotypic traits in pathogenesis, especially when combined with relevant animal model systems. Furthermore, the structural analysis of important virulence factors and surface antigens may allow the prediction of antigenic domains suitable for the development of new vaccines.

The 38th Mosbacher Colloquium focuses on the molecular basis of viral and microbial pathogenesis. The virology part begins with the well-studied plant viroids. The unusual structure of their genome, as well as knowledge about their replication and pathogenicity, are presented. The molecular basis of virus variability, the role of viral glycoproteins as determinants of pathogenicity and the significance of viral genes in defining the host range of influenza virus infections represent further important topics discussed. Other basic problems for the understanding of viral infections are DNA amplification by viruses, the control of viral promoters and bio-pathogenic aspects of persistent infections that impede neurofunctions.

Much progress has recently been achieved in understanding the molecular basis of virulence of those bacteria that colonize cell surfaces. In particular, the genes for a large number of bacterial toxins, for specific surface structures involved in adhesion and for antiphagocytic activity have been studied in great detail. Significant contributions are provided here on the genetics of Shiga-like toxin, pertussis toxin, cytolysins as well as tetanus and botulinum toxins. The antigenic variations in *Neisseria gonorrhoeae* and the molecular architecture of the streptococcal M proteins are exciting examples of the importance of bacterial surface structures in evading the host defense. The recent progress in understanding the mechanisms involved in invasion and survival of intracellular bacteria is exemplified in elegant genetic and biochemical studies on *Shigella flexneri*.

The final part of this colloquium addresses modern trends in the molecular biology of parasites, the oncology of one-gene-transducing viruses and the important question of how host genes control the susceptibility to a virus. The 38th Mosbacher Colloquium thus provides an excellent insight into a fascinating field of modern biochemical and medical research.

Acknowledgements

The organizers are grateful to the Gesellschaft für Biologische Chemie and its Chairman Prof. Dr. E.-L. Winnacker for their active support of this colloquium. Special thanks are due to Prof. Dr. H. Gibian and Dr. E. Truscheit and all their co-workers for technical help and to all individuals and institutions, in particular the Deutsche Forschungsgemeinschaft, for providing funds and contributing towards making this colloquium a success.

September 1987 R. Rott and W. Goebel

Contents

Contributors

You will find the addresses at the beginning of the respective contribution

Correlation Between Viroid Structure and Pathogenicity

H. L. Sänger and B. Haas[1]

Introduction

Viroids are the smallest and structurally best-characterized disease agents presently known. Thus far, they have only been found in higher plants, where they cause several economically important diseases. Viroids are unencapsidated single-stranded covalently closed circular RNA molecules with a chain length of about 240 to 380 nucleotides, depending on the viroid "species". The sequence of about a dozen different viroid species and numerous "isolates" or "variants" thereof is known. Detailed biochemical and biophysical studies have shown that all known viroids follow a common principle of structure and dynamics. They exist in their native state as unbranched double-stranded rod-like structures in which short base-paired regions are alternating with small internal and bulge loops (for review, see Diener 1983; Sänger 1984; Riesner and Gross 1985).

In contrast to the wealth of information on the structure of viroids, little is known about the biological functions and interactions with the host cell, which is mainly due to the experimental limitations and difficulties inherent in plant systems. One of the still unresolved problems in viroid research centers around the question how such relatively small RNA molecules with their limited genetic information cause disease in higher plants.

Several lines of evidence suggest that viroids are not translated into proteins and peptides (Davies et al. 1974; Hall et al. 1974; Semancik et al. 1977). The observed accumulation of certain proteins in tomato plants infected with PSTV (Zaitlin and Hariharasubramanian 1972; Camacho and Sänger 1982, 1984; Galindo et al. 1984) and in *Gynura aurantiaca* infected with CEV; Flores et al. 1978; Conejero et al. 1979) is not a viroid-specific response but a general pathophysiological reaction of the host to infection and also induced by conventional viruses and fungi (Camacho and Sänger 1982a,b, 1984). Therefore, there is general agreement that viroid pathogenicity can only be exerted by the viroid RNA itself.

There are two possible ways how to approach the problem of viroid pathogenicity experimentally, namely from the host plant with all the complexity of its metabolism and its reactions or from the viroid molecule with its fully elucidated and comparatively simple RNA structure. Therefore it is not surprising that in the past the corresponding studies have concentrated preferentially on trying to correlate the sturcture of the viroids with their pathogenicity. At present newly developed methods of recombinant DNA technology provide very promising perspectives in that transgenic plants can be produced, so that in the near future the pathogenic interaction between the viroids and their plant hosts can also be studied appropriately at the level of

[1] Max-Planck-Institut für Biochemie, Abteilung Viroidforschung, D-8033 Martinsried, FRG

38. Colloquium Mosbach 1987
Molecular Basis of Viral and Microbial Pathogenesis

the host cell. The purpose of this short overview is to summarize the present knowledge on the correlation between the pathogenicity (virulence) of viroids, their molecular structure and the possible host targets of these pathogens.

Structure Pathogenicity of Potato Spindle Tuber Viroid (PSTV)

The potato spindle tuber viroid (PSTV) represents the prototype viroid because it was the first discovered member of this novel class of plant pathogens (Diener 1971). The primary and secondary structure of PSTV with its 359 nucleotides are known (Gross et al. 1978) and its structural (Sänger et al. 1976) and thermodynamic (Riesner et al. 1979) properties have been characterized in detail. Different PSTV field isolates produce disease in tomato plants characterized by mild, intermediate and severe symptoms and even by the death of the infected plants (Fernow 1967; Dickson et al. 1979; Gross et al. 1981; Sänger 1982).

Since RNA is the only component of viroids, differences in the virulence of these isolates can be based only on differences in the sequence of their RNA. For a definite assessment of the correlation between viroid structure and viroid pathogenicity we sequenced a series of individual PSTV field isolates of different virulence (Fig. 1). The sequence of their RNA was established at the level of their cDNA which was synthesized by primer extension of the PSTV RNA with reverse transcriptase using three PSTV-specific DNA primers. The PSTV cDNA's were then sequenced with the Maxam-Gilbert technique. This approach was possible because in all cases the majority of the PSTV molecules of a given isolate was represented by one RNA species. Therefore, we assume that the population of these predominating RNA molecules is responsible for the phenotype of the disease observed.

The sequence analysis of seven PSTV isolates revealed (Schnölzer et al. 1985) that substitutions, insertions and deletions of nucleotides are found at three distinct sites of the primary structure of the PSTV RNA molecule (Fig. 2). As compared with the PSTV type strain (PSTV DI) they are located between nucleotides 45-50, 119-122 and 308-318, respectively. Regarding the secondary structure model, nucleotides 45-50 and 308-318 are partly complementary and positioned opposite to each other in the upper and lower strand in the left hand part of the rod-shaped molecule, whereas the region between nucleotides 119 and 122 is located in its upper right hand part. In all those PSTV isolates where, due to the mutations in the left hand part of the molecule, one nucleotide is deleted (DI, KF 440-1, KF 440-2), this deletion is always compensated for by the substitution of one uridine by two adenosines at position 120/121 in the right hand part so that the total number of the 359 nucleotides in these PSTV isolates is strictly maintained.

Nucleotide Changes and Secondary Structure of PSTV

The influence of the observed nucleotide changes on the secondary structure of the corresponding PSTV isolates was evaluated by computer-aided thermodynamic calculations (Schnölzer et al. 1985). This analysis allows the determination of the influence of small sequence differences on the secondary structure of the PSTV RNA molecule. It was found that the observed mutations induce only local structural changes at the site where they occur and that they do not influence the overall secondary structure of PSTV. Moreover, only the changes in the left hand part of the molecules (Fig. 2) seem to be functionally important for the viru-

Control KF6 DI HS KF440

Fig. 1. The virulence of PSTV for tomato. Tomato plants (cultivar Rutgers) are shown 8 weeks after mechanical inoculation by leaf rubbing with tissue homogenates from tomato plants infected with different PSTV field isolates. *From left to right* uninfected healthy control plant, plants infected with isolates inducing mild (KF 6), intermediate (DI), severe (HS) and lethal (KF 440) symptoms of disease, respectively

lence of PSTV, because due to those changes the thermodynamic stability of a distinct region, the so-called pre-melting (PM) loop 1 region, becomes different in the isolates of different virulence.

Previous thermodynamic studies on the PSTV type strain (PSTV-DI) have shown that its rod-like shape is maintained up to temperatures close to a rather sharp melting transition (Riesner et al. 1983; Steger et al. 1984). However, there are three "pre-melting (PM) regions" 1, 2 and 3) in the molecule, which are already formed below this melting temperature (Fig. 3a). PM loop 1 is formed between nucleotides guanosine 49-adenosine 60 in the upper and between nucleotides uridine 300-uridine 312 in the lower strand and comprises a part of the region of mutation (RM) where practically all of the nucleotide changes occur. The upper strand of PM loop 1 is characterized by a polypurine sequence, whereas the lower strand contains the conserved sequence 3'UCUAU-CUC-5' (Fig. 3b). Under the presumption that the observed mutations modulate the virulence of the PSTV isolates by changing the thermodynamic behaviour of the region of PM loop 1 we denote the region of PM loop 1 together with the region of mutation the "virulence-modulating (VM region)" of PSTV (Fig. 3a,b). The free energy to form PM loop 1 in the structures shown in Fig. 3b was calculated for 25°C and 1 M NaCl. It was found to be 4.51, 2.53, 0.75 and 0.43 kcal/mol for the mild, intermediate, severe and lethal isolate, respectively. PM loop 2 which is formed between nucleotides guanosine 73-cytosine 93 and guanosine 266-cytosine 285 in the upper and lower strand, respectively, comprises parts of the conserved central region of the PSTV molecule (compare Fig. 3a with 6a) and is therefore found in most viroid species. Its strict conservation indicates that this region is of functional importance and that it might be essential for viroid replication. PM loop 3 of PSTV is located between guanosine 134-guanosine 143 in the upper strand and cytosine 218-cytosine 228 in the lower strand and conserved in all PSTV isolates. Such a pre-melting region exists

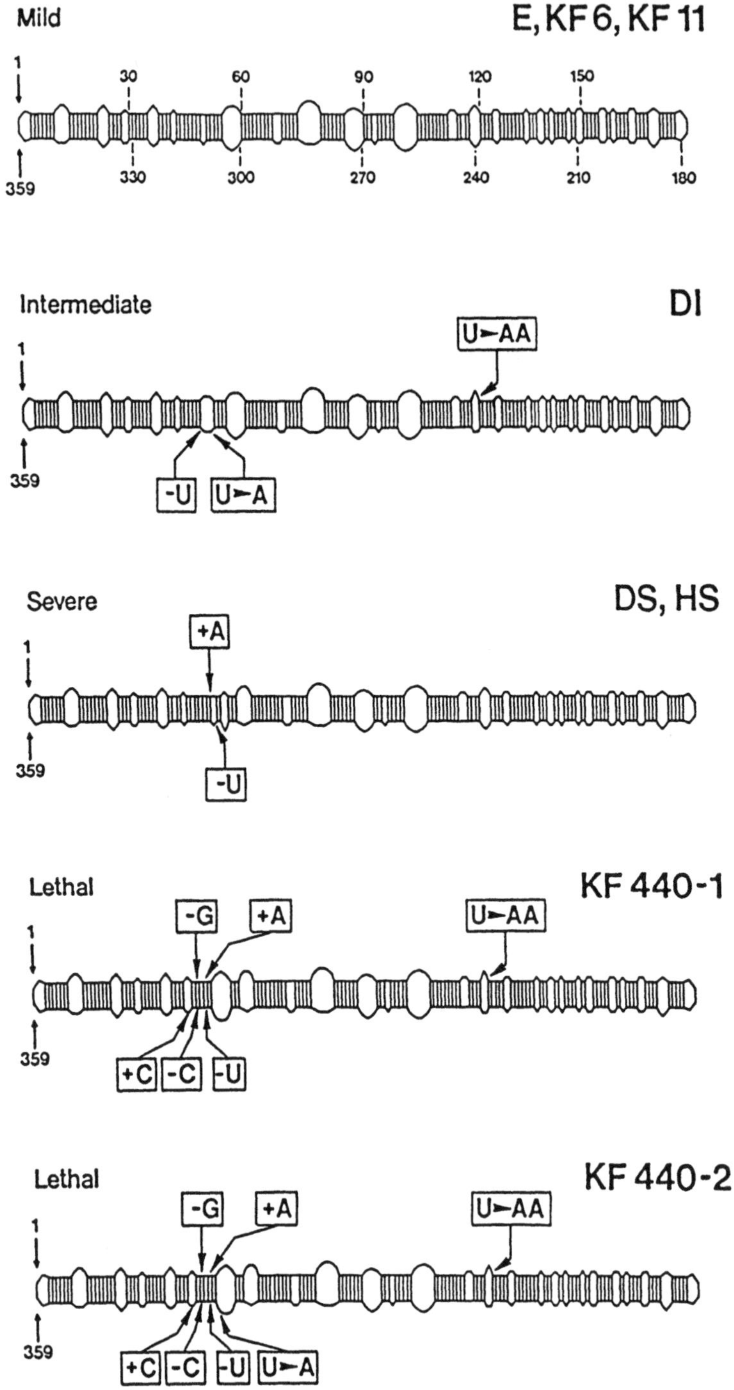

Fig. 2. The location of the nucleotide changes at the secondary structure model of the PSTV isolates causing mild, intermediate, severe and lethal disease symptoms in tomato (cv. Rutgers) as shown in Fig. 1. The changes are indicated in *boxes* and refer to the mild isolates as standard

at practically the same position in most other viroids although the corresponding sequence is not conserved. Regarding the importance of all these premelting regions one could postulate that they provide the basis for specific structural rearrangements of the viroid RNA and contribute to its functional flexibility.

The compensating nucleotide change at the right hand side of the PSTV RNA where one uridine is replaced by two adenosines (Fig. 2) slightly increases the stability of this region via extending the helix by one base-pair and by concurrently decreasing the adjacent internal loop at the left hand side of this short helix by one nucleotide.

Concentration, Melting Behavior and Virulence of PSTV

Differences in PSTV virulence could be primarily based on the amount of PSTV RNA accumulating in the host plant tissue after infection. The observed sequence differences could, for example, influence viroid synthesis and modulate disease in this way. Therefore, we analyzed at different times after inoculation the PSTV RNA content in comparable newly developing leaves of tomato plants infected with the various PSTV isolates. However, no significant differences in the PSTV RNA content were found between the isolates of different virulence when the corresponding RNA samples were separated in bi-directional runs on polyacrylamide gels (Schumacher et al 1983a) and compared after visualization by silver staining. Consequently, another mechanism of PSTV disease modulation must operate.

Under the assumption that the observed nucleotide changes might modulate virulence via altering the structural features of the PSTV RNA molecule, the influence of these changes on the thermodynamic properties of the VM region, as reflected in the melting behavior of PM loop 1, was analyzed theoretically with the aid of a computer program as described by Schnölzer et al. (1985). It was found that PM loop 1 within the VM region of the mild isolates exhibits the highest melting temperatures of 74°C and that with increasing virulence of the isolates the values of their T_m decrease down to 50°C in the case of the lethal isolates. However, these differences in the local melting behavior within the VM region do not influence the overall melting behavior of the entire PSTV molecule. The experimentally determinded T_m value of the main transition of PSTV as extrapolated to 1 M NaCl is 77°C and the theoretically calculated one is 78°C (Steger et al. 1984).

The presence of concatameric viroid-specific (-)RNA molecules of different size in the infected plant as integral forms of viroid replication (Rohde and Sänger 1981; Branch et al. 1981; Spiesmacher et al. 1983) prompted us to analyze also the corresponding region of this (-) RNA of the various PSTV isolates. However, neither any correlation between their virulence and the thermodynamic stability of a putative VM region in their (-)RNA was found, nor were any other features detectable from which a possible mechanism for PSTV virulence could be derived. This result and the fact that PSTV(-)RNA molecules are present in the host cell only in extremely low concentrations strongly suggest that the monomeric PSTV(+)RNA is primarily responsible for the pathogenic action of viroids. It is still unclear, however, whether the disease is incited by the mature monomeric circular PSTV progeny which accumulates up to 10,000 molecules in the nucleolus (Schumacher et al. 1983b). Alternatively, the PSTV(+)RNA could exert its pathogenic action in its immature state. One could visualize that the interaction takes place between the host cell target and the nascent RNA chain prior to its final processing and maturation. In this case the RNA could induce

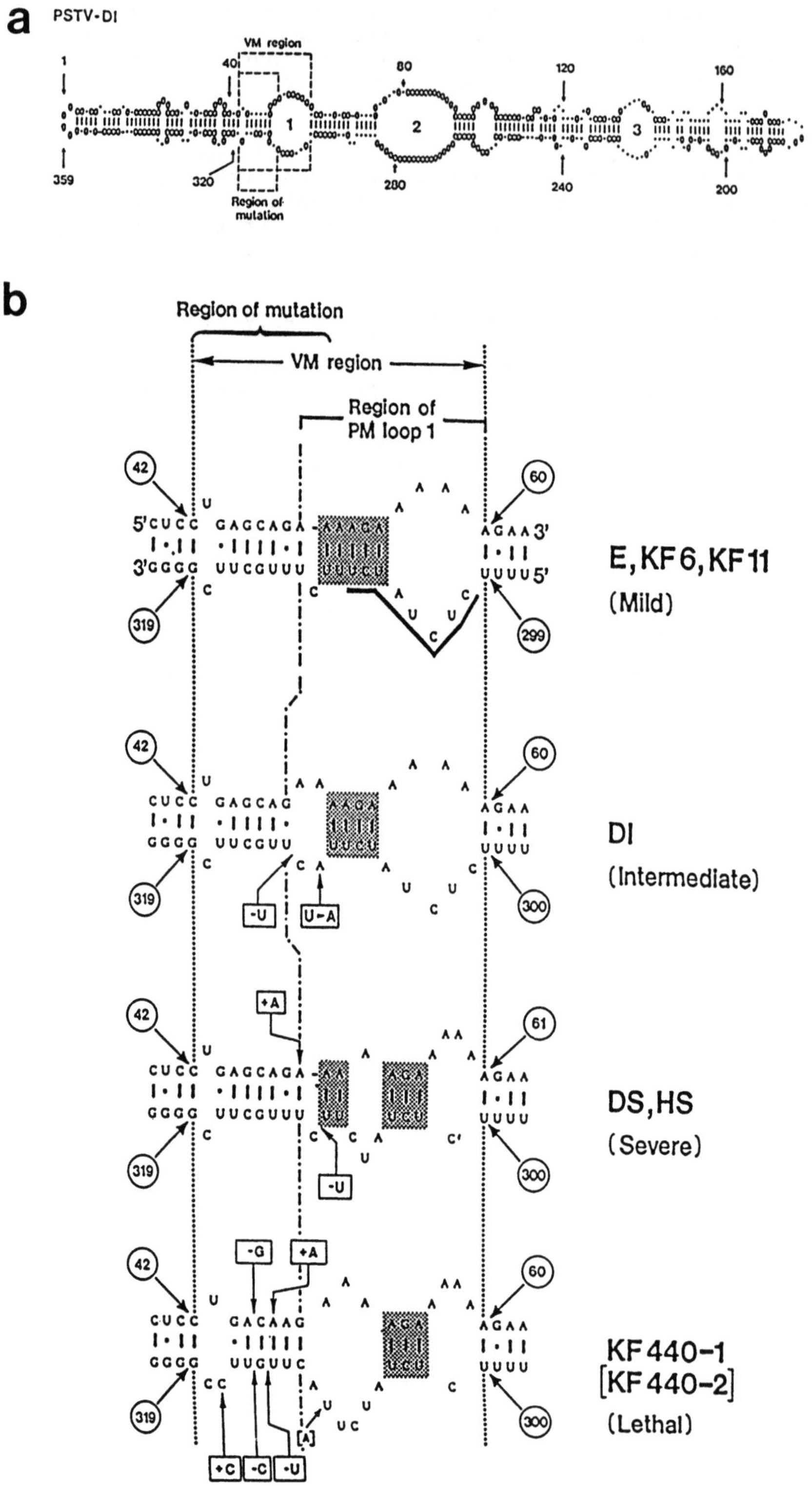
a
PSTV-DI
VM region
Region of mutation
1
40
80
120
160
359
320
280
240
200
b
Region of mutation
VM region
Region of PM loop 1
E, KF6, KF11
(Mild)
DI
(Intermediate)
DS, HS
(Severe)
KF440-1
[KF440-2]
(Lethal)
-U
U→A
+A
-G
+C
-C
[A]

a series of structural transitions while growing, and assume a configuration in which a certain region of the viroid RNA is transiently exposed and amenable for interactions with the cellular target.

Structure and Pathogenicity of Citrus Exocortis Viroid (CEV)

The occurrence of mutations primarily in one distinct domain of its rod-shaped molecule seems to be a special feature of PSTV. A different situation exists in the citrus exocortis viroid (CEV) where nucleotide changes are distributed all over the molecule with an accumulation at distinct regions, one at the left-hand and another at the right-hand part of the molecule. The sequence analysis of five different field isolates has identified 15 sequence variants in these isolates which were grouped into Class A and Class B variants. As compared with the CEV type strain (CEV-A) the class A variants show nucleotide changes primarily in the left-hand part of the molecules and they produce severe disease symptoms in tomato. The class B variants exhibit changes in both parts of the molecule and induce only a mild disease. Therefore the two regions where the nucleotide changes predominate were called P_L and P_R, the left and right pathogenicity region of CEV (Fig. 4a). It was, however, not clear whether one or both of these domains were responsible for the variation in pathogenicity (Visvader and Symons 1985). Similar class A and B CEV isolates (CEV-CG1 and CEV-LG1) producing severe and mild symptoms on *Gynura aurantiaca*, respectively, were sequenced in our laboratory (Fig. 5).

From these sequencing data alone the influence of the observed nucleotide changes on the virulence of CEV could hardly be assessed. Therefore, Visvader and Symons (1986) have constructed viroid cDNA mutants and used those for infectivity studies. The molecular cloning of a severe class A [CEV-A(2)] and a mild class B [CEV-DE 30(a)] sequence variant of CEV (Fig. 4a) allowed the construction of chimeric cDNA clones in which the P_L and P_R domains from both variants were arranged in the two possible ways in order to study the relative effects of the two domains on viroid pathogenicity (Fig. 4b). One chimeric construct designated S_L/M_R, contained the left hand part of the severe variant, CEV-A(2), adjoined through the BamH I and Hind III site present in their conserved central region to the right-hand part of a mild variant, CEV-DE30(a). The reverse chimera, M_L/S_R, was constructed using the same parental clones.

When a mixture of plasmid DNA and RNA transcripts from the chimeric M_L/S_R clone was inoculated onto tomato plants, mild symptoms were induced in all plants in four separate experiments. Upon further passage in tomato this property remained stable and mild symptoms were further produced. However, the S_L/M_R clone and its RNA transcripts elicited both mild and severe symptoms, in which case the ratio of mild to se-

Fig. 3a,b. The virulence-modulating (VM) region of PSTV. a Location of the VM region in the PSTV type strain (PSTV-DI) in its secondary structure model as calculated for 70°C, which is 8°C below the main melting point of the viroid. The nucleotides represented by *open circles* are identical in the three viroid "species", PSTV, CEV and CSV, the varying ones are represented by *points*. In contrast to the native structure, the pre-melting structure of PSTV exhibits three larger inner PM loops (indicated by *numbers* within the structure) of which PM loop 1 and 2 are characterized by nucleotides conserved also in CEV and CSV. b The conserved sequence 3'-UCUA-UCUC-5' present in the lower strand of the region of PM loop 1 of all isolates is underlined by a *black bar* in the mild isolates. The shadowed base-paired region dissipates when PM loop 1 is formed

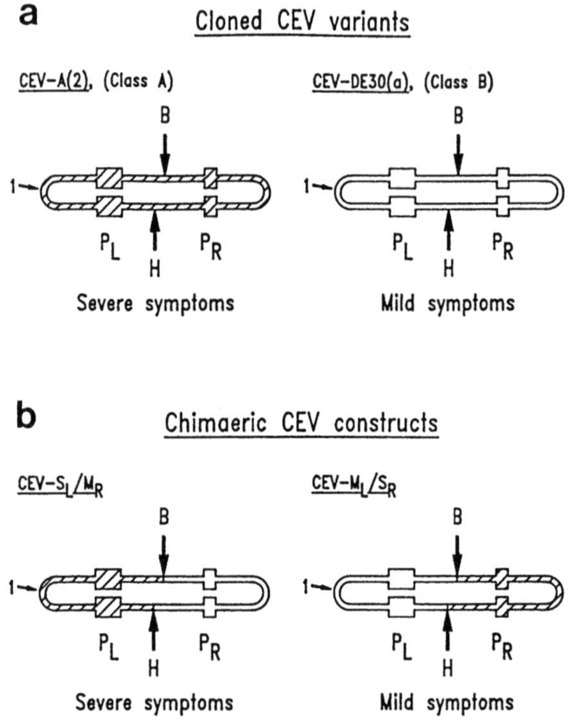

Fig. 4a,b. Demonstration of virulence modulation in CEV by its P_L region. a Schematic representation of the two cloned parental CEV variants. b Two chimeric CEV constructs produced from the cloned parental variants. The relative position of the putative pathogenicity regions P_L and P_R, of the BamH I(B) and Hind III (H) restriction sites used for cloning and constructing the chimara, and the virulence of the four CEV forms for tomato are indicated

vere was dependent on the amount of inoculum used per tomato seedling. Northern hybridization analysis of nucleic acid extracts prepared from pooled plants of a single experiment (using only plants with severe symptoms in the case of chimera S_L/M_R) showed that the two chimeric clones gave levels of viroid a little lower than that of the clone of the mild variant CEV-DE30(a), but higher than that of the severe variant CEV-A(2). From these bioassays it can be concluded that the P_L domain of CEV plays a direct role in determining whether the symptoms are mild or severe, while the P_R domain does not. Therefore, the P_L domain of CEV corresponds to the VM region of PSTV and incites the disease most probably by interaction with the same host target.

Of particular interest is the finding that the severity of symptoms observed after inoculation of tomato seedlings with the chimeric clone S_L/M_R varies with the concentration of nucleic acid used in the inoculum. With highly concentrated inocula the number of plants with severe symptoms increases. Moreover, the viroid concentrations in plants with severe symptoms were 9-20 times higher than those in extracts of plants showing mild symptoms. Hence, at least in the case of infection with the chimeric variants S_L/M_R, symptom expression seemed to be also dependent on the level of viroid which accumulated in the plant. These experiments indicate that in this variant the efficiency of certain steps of the infection or replication process might have been affected, resulting in the observed variation in the level of progeny viroid found between different plants. One could argue that the reduced infectivity of the S_L/M_R chimeric clone might be due to the disruption of long-range interactions that have been postulated to occur between distant parts of the viroid molecule (Riesner et al. 1979, 1981; Riesner and Gross 1985).

The variation of symptom expression with viroid concentration is surprising in view of previous results of Visvader ans Symons (1985),

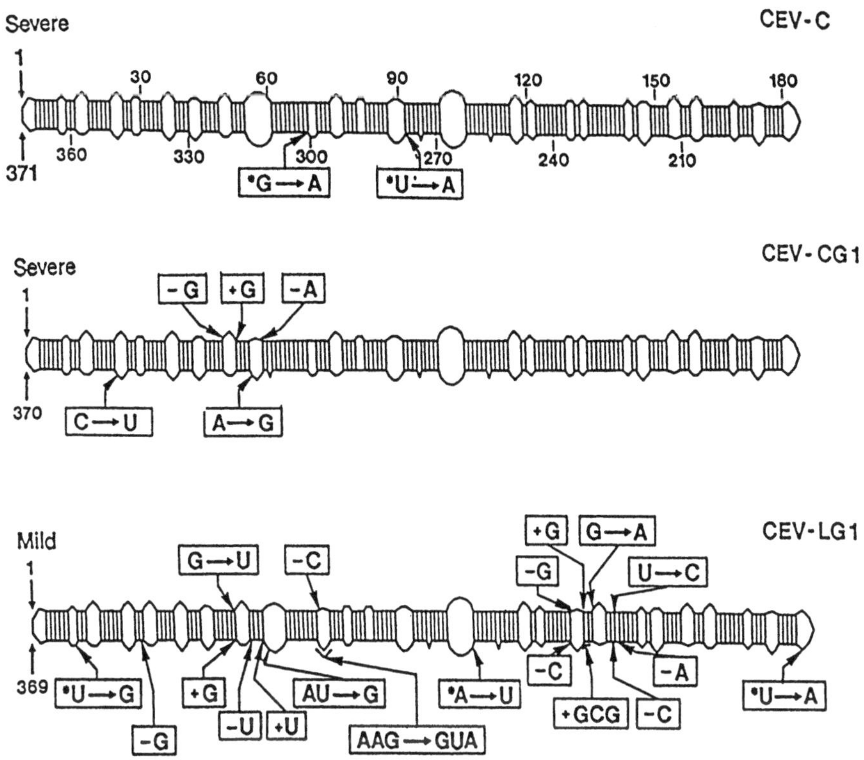

Fig. 5. Sequence differences between three isolates of citrus exocortis viroid (CEV) which differ in their virulence for *Gynura aurantiaca*. The nucleotide changes are indicated in boxes and refer to the type strain CEV-C. The nucleotide changes in CEV-CG1, a class A variant, are found to accumulate only in a distinct region in the left hand part whereas in CEV-LG1, a class B variant, they are mainly located in two distinct regions, one in the left and the other in the right hand part of the molecule

which indicated that differences in symptom expression between naturally occurring mild and severe CEV isolates (each containing a mixture of sequence variants) were independent of the viroid level in infected plants. These findings demonstrate that infectivity trials with in vitro-constructed mutants should be interpreted with caution. They also emphasize the importance of multiple experiments in which inocula of different concentrations should be used.

VM Region in Other Viroids

The nucleotide sequence of nine viroid "species", namely PSTV (Gross et al. 1978), CEV (Visvader and Symons 1983; Visvader et al. 1982; Gross et al. 1982), CSV (Gross et al. 1982; Haseloff and Symons 1981), ASBV (Symons 1981), CCCV (Haseloff et al. 1982), HSV (Ohno et al. 1983), CPFV (Sano et al. 1984), TPMV (Kiefer et al. 1983) and TASV (Kiefer et al. 1983), is presently known. Among these the sequences of CEV, CSV, TPMV and TASV share about 60-80% similarity with the one of PSTV, so that all these viroids have been allocated to the "PSTV group" (Kiefer et al. 1983). HSV and CPFV, which are very closely related to each other, share only 46% of their sequence with PSTV and can thus be considered as members of a separate group, namely the "HSV group". The host range of the viroids of these two groups, although restricted to dicotyledonous plants, is largely overlapping (Diener 1979). CCCV (Haseloff et al. 1982) and avocado sunblotch viroid (ASBV) (Symons 1981), represent individual unrelated viroid "species" and exhibit a sequence similarity to PSTV of only 56 and 39%, respectively. Their host range is extremely limited, as CCCV is replicated only in the monocotyledonous coconut palm (*Cocos nucifera*) and in some closely related palm species (Randles et al. 1980; Imperial et al. 1985), whereas ASBV can only be transmitted to avocado (*Persea americana*) and cinnamon (*Cinnamomum zeylanicum*) which are both members of the dicotylodonous family Lauraceae (Da Graca and Van Vuuren 1980).

With the exception of ASBV and CCCV a domain similar to the VM region with its premelting loop 1 region of PSTV can be detected in all viroids, at the same topological position of their rod-shaped secondary structure as shown for PSTV in Fig. 6a. It exhibits a high degree of sequence similarity in particular to the region of PM loop 1 of PSTV with its purine-rich sequence in the upper and the characteristic palindromic hepta nucleotide sequence 3'-UCUAUCU-5' in the lower strand, respectively. In ASBV this similarity to the VM region is only partial and restricted to the lower strand of its RNA structure whereas in CCCV no similarity is detectable at all. In the light of our model for the pathogenicity of PSTV, this is not too surprising because CCCV and ASBV also cause a completely different type of disease in their hosts, which is characterized by yellowing type symptoms in contrast to the growth retardation and leaf malformation incited by the other viroids. Moreover, CCCV and ASBV are evidently not associated with the cell nucleus like PSTV (Spiesmacher et al. 1983, 1985; Schumacher et al. 1983b) and the others, but are found predominantly with the cytoplasm and its reticulum (CCCV) (Randles et al. 1976) or in the chloroplasts (ASBV) (Mohamed and Thomas 1980). From all these differences one could expect that the mechanism of pathogenesis of these two viroids (and probably also their site and mode of replication (Spiesmacher et al. 1985) might differ from the one we have developed here for PSTV and its relatives. Since no comparative studies are available, as yet, on the correlation between the thermodynamic parameters and the virulence of different isolates of the other members of the PSTV and HSV group, it is still unknown whether or not their virulence is also specified and modulated by this domain as we propose for PSTV and CEV.

Models Proposed for Viroid Pathogenicity

It has been postulated that viroid RNA might act as an aberrant kind of autoinducing "regulatory RNA" (Diener 1971, 1977, 1978; Semancik and Weathers 1972; Robertson and Dickson 1974; Reanney 1975; Dickson and Robertson 1976; Dickson 1979; Zimmern 1982) and cause disease by disregulating the expression of host genes. From the disease symptoms induced by viroid infections such as growth retardation and stunting of plants, combined with curling and malformation of their leaves, it has been concluded that viroids induce disturbances in the metabolism of plant growth hormones (Rodriguez et al. 1978; Semancik 1979; Duran-Vila and Semancik 1982). From the total dependence of viroid replication on host enzymes it has been surmised that viroid RNA may incite disease by usurping the corresponding DNA-dependent RNA polymerases for its "selfish" replication (Rackwitz et al. 1981). However, the absence of a cytopathogenic effect in rapidly dividing cells of tomato and potato suspension cultures which continuously synthesize and accumulate PSTV RNA molecules to high concentrations (Mühlbach and Sänger 1981; Mühlbach et al. 1983) would preclude that the pathogenic action of PSTV is simply due to its presumed capability of simply "titrating out" the RNA polymerases of the host cell. Therefore, it is rather unlikely that the observed specific binding of RNA polymerases II to viroid molecules (Goodman et al. 1984), and the rate of viroid replication per se are responsible for the induction and modulation of disease. On the basis of certain sequence homologies it has furthermore been argued that the viroid RNA and/or its complement interferes with the components and mechanisms involved in pre-mRNA splicing (Gross et al. 1882; Diener 1981; Dickson 1979), thereby perturbing normal mRNA processing. Of particular interest in this context is the model of Dickson (1981), which was the only one which took into account the sequence differences between a mild and intermediate PSTV isolate as known at that time. If, however, our additional sequencing data are incorporated in this model, the proposed correlation between the severity of the disease and the interaction of each viroid to the splice function cannot be maintained. All the other hypotheses correlate the pathogenic action of viroids and their ability to modulate disease through the interaction of the U1 RNA and the mRNA splice junction with other domains of the PSTV molecule than the region where the virulence-related nucleotide changes have actually been detected. On the basis of our and other data this inconsistency has been discussed in detail by Flores (1984).

On the basis of the previously published sequences of PSTV isolates a conformational model for viroid pathogenesis has been proposed (Flores 1984) which differs from our model. Flores correlated the pathogenicity of PSTV isolates with the alteration of the "amplitude" (i.e., the size) and the number of "central" (i.e., internal) loops in a segment of the rod-like structure of the PSTV RNA molecule between nucleotides 44-55 in the upper and nucleotides 306-317 in the lower strand. However, the proposed structures and conformational changes show that a thermodynamic analysis is completely lacking and that an erroneous sequence (PSTV-Scotland) has been used which led to an erroneous assignment of intermediate pathogenicity of this actually severe to lethal isolate. Moreover, this conformational model does not provide any mechanism for viroid pathogenesis and it does not allow to predict the virulence of a given isolate.

Most recently, it has been argued that viroids might exert their pathogenic effect by interfering with the normal pre-rRNA processing in plant nucleoli. This interference is considered to consist in a strong binding of viroid RNA to the internal transcribed spacer 2 (ITS2), which is an intron-like sequence between 5.8S rRNA and 25S RNA that is excised during pre-rRNA processing. Thus, the ITS2 would become inaccessible to U3 small nuclear RNA which is thought to be involved in the proper excision of ITS2 (Jakab et al. 1986). However, unpublished sequencing work in the our laboratory has shown that the ITS of the cytoplasmic rRNA genes of tomato exhibit no sequence complementarity whatsoever with PSTV.

As has been discussed in detail (Diener 1983; Flores 1984), none of these models, including the most recent one of Jakab et al. (1986), can readily explain in molecular terms how a viroid specifies and modulates disease and why it is pathogenic in some hosts yet harmless in others.

It should be stressed in this context that the different response of the various species and cultivars of tomato (Sänger and Ramm 1975; Mühlbach et al. 1977) and potato (Diener 1979) to PSTV infection is genetically determined and therefore requires species- and cultivar-dependent differences in the viroid-specific host target. In case the target is located at the host plant genome one must assume that it cannot be one of the strictly conserved elements of the host DNA but rather one which varies in sequence and number. Such elements have, in fact, been detected in plants at the genomic level in the multigene family of the cytoplasmic rRNA of wheat. The total number of these genes including their non-transcribed regions with their repetitive promoter/enhancer elements varies remarkably in various wheat cultivars and related species (R.B. Flavell, unpubl. results). Under the assumption that similar differences also exist in the organization of the genome of the different viroid host plants and their cultivars, one could expect that the viroid RNA interacts with those variable elements in a specific and yet differentiated way.

Model for the Pathogenic Action of PSTV and its Modulation

We have found that the virulence of the PSTV isolates is specified by the nucleotides in their so-called region of mutation (RM). We furthermore found a correlation between the severity of these isolates and the ease with which the region of PM loop 1 can be melted. Therefore we wondered how this melting property modulates disease. The most plausible assumption would be that the nucleotides of the VM region of PSTV interact in some way with a host factor(s), thus disturbing its normal cellular function. Since the thermodynamic stability of the region of PM loop 1 decreases with increasing virulence, one can conclude that concurrently the binding of the host factor(s) becomes more probably, which would aggravate the resulting disease. The observed correlation would also indicate that the host factor(s) is apparently specific for a single-stranded part of the VM region. Moreover, the binding site of the host factor(s) should comprise in all PSTV isolates the same nucleotides in the conserved nucleotide sequence of PM loop 1. Under these premises, the observed nucleotide changes in the region of mutation would specify and modulate PSTV virulence by altering the accessibility of the target sequence for the host factor(s) the PSTV RNA molecule. It follows that the nucleotides of the upper or lower strand of PM loop 1 are most probably the only site where the binding of the host factor(s) can take place. However, our thermodynamic analysis of the PSTV isolates does not allow us to decide to what extent the nucleotides of the double-stranded regions adjacent to the left and right hand side of PM loop 1 are also involved in this interaction. Moreover, the nature of the putative host factor(s) has not been determined as yet.

Assuming that the VM region is recognized as target by a cellular protein, this could bind to the nucleotides of the lower strand of PM loop 1 in a stable tertiary strcture which would be favored by the stacking of the polypurine stretch in the corresponding upper strand. In cases where a nucleic acid would base-pair with the nucleotides of PM loop 1, the resulting binding energy would not be sufficient for binding this nucleic acid factor. However, the extensive sequence similarity present at both sides of the region of PM loop 1 in all viroids except CCCV and ASBV would suggest that such a nucleic acid might be complementary to a much longer sequence in this region and hence exhibit a higher affinity for the viroid molecule. Such a complex could, of course, become more stabilized by an additional host protein which binds to the nucleic acid complex.

The interaction between PSTV RNA and a putative host factor(s) could be influenced not only by parameters which determine the state of equilibrium but also by the activation energy which defines the kinetics of how the equilibrium is reached. Therefore, the VM region with or even without the bound host factor(s) could exert its virulence-modulating function in that it initiates a structural transition of the viroid RNA molecule. Other viroid sequences might then become exposed in structures such as bulge loops, distorted helices or hairpins, all of which have been shown to be responsible for the recognition and binding of proteins by RNA molecules (Gralla et al. 1974; Garret et al. 1981; Krug et al. 1982; Carey et al. 1983). Although the actual structure of viroids in vivo is unknown, the unique thermal melting behavior of these rod-shaped RNA molecules with their various transient metastable and intermediate structures (Riesner et al. 1983; Steger et al. 1984) suggests that such structural rearrangements are not only potentially possible but they probably reflect an essential feature of functional importance not only for their replicability and processibility, but also for disease induction by this unique class of pathogenic small circular RNA's.

The predominant occurrence of mutations in one part of PSTV indicates that the corresponding nucleotides are not essential for the "survival" of viroids. In fact, our model implies that pathogenicity and replicability are controlled by different domains of their RNA genome. From the strict conservation of the central region in all viroids and from the requirement of its presence for the infectivity of cloned PSTV DNA (Tabler and Sänger 1984, 1985) one must infer that the central domain is one of the essential structural prerequisites for the transcription or the subsequent processing of viroids (Tsagris et al. 1987).

Implications of the Pathogenicity Model

Our model is able to account for the modulation of PSTV-incited disease symptoms and for several other phenomena specific for the development of disease in PSTV-infected plants which could not be interpreted so far in a plausible way in molecular terms. (1) The increase in the severity of the disease symptoms with the increasing ambient temperature at which tomato plants are held after PSTV infection can now be at least partly related to its direct effect on the stability of the PM loop 1 region of the PSTV molecule. This region becomes destabilized with increasing temperature, which increase in total the binding affinity of the viroid molecules for the putative factor(s) of the host and thus its pathogenic response. It should be stressed, however, that the effect of temperature on symptom expression is certainly much more complex because any increase in temperature will also increase PSTV synthesis and accumulation (Sänger and Ramm 1975). In addition, temperature also affects host plant metabolism directly and may thus influence disease development by altering the reactivity of the host. (2) The cultivar-dependent disease response of potato (Diener 1979) and tomato (Mühlbach and Sänger 1977; Mühlbach et al. 1977) to PSTV infection can now be attributed to genetically determined differences in the number of the host factor(s) or its substrate(s) in these cultivars. This would lead to quantitative differences in the competitive interaction between the RNA molecules of the PSTV isolate and these host cell components. (3) In all nonresponding plants in which viroids are well replicated without causing disease, the putative host factor(s) responsible for the induction of viroid pathogenesis could either be present in great excess, or be completely absent, or lack the affinity for binding to the VM region of the viroid in question.

The Host Plant in the Pathogenic Interaction

Our pathogenic model has been derived from the sequence and the thermodynamic analysis of the RNA genome of different PSTV field isolates. Only circumstantial evidence is available so far for the nature of the host target which is required for the pathogenic interaction. One could postulate that it takes place at the site of PSTV (+) RNA synthesis and accumulation. It has been shown that PSTV is replicated inside the nucleus and that the DNA-dependent RNA polymerase II of the host is involved in the strand-specific synthesis of PSTV (+) and (-) RNA, respectively (Spiesmacher et al. 1983, 1985). The location of this RNA polymerase suggests that not only the PSTV (-) RNA but also the linear oligomeric PSTV (+) RNA transcripts are synthetized and further processed in the nucleoplasm. However, the resulting viroid progeny, i.e., the circularized monomeric mature, rod-shaped PSTV (+) RNA molecules accumulate in the nucleoli (Schumacher et al. 1983b) where they are apparently complexed with histones and other nucleolar proteins (Wolff et al. 1985). These proteins and the genomic ribosomal DNA constitute the nucleosomal structure. Therefore, the question arises whether or not PSTV and presumably also its relatives exert their pathogenic action inside the nucleolus where they could interfere with the transcription of the cytoplasmic rRNAs. It is tempting to speculate that promoter and enhancer elements at the genomic rDNA and/or the corresponding regulatory proteins are the host cell targets with which the viroid RNA can interact.

Tomato 7S RNA as a Possible Cellular Target for PSTV

In search for cellular RNA molecules which could be the putative target with which viroids might interact and thus cause disease, we have sequenced the 7S RNA of tomato (*Lycopersicon esculentum*, cultivar Rentita). We proposed a secondary structure model which is based on a combination of computer modelling and data obtained from nuclease digestion experiments. Next of tRNA and 5S RNA, 7S RNA represents one of those plant RNA species of low molecular weight whose sequence and biological function have not been determined as yet.

Our sequence analysis has shown that the 7S RNA of tomato has a chain length of 302 nucleotides considering three uridine residues out of a variable number at the 3'-terminus. The availability of the total sequence of the 7S RNA from tomato as the most widely used host plant in viroid research allowed us to compare this sequence with the sequence of PSTV. The result of this comparison shows (Fig. 6) that a significant sequence complementarity exists between nucleotides 24-100 at the 5'-terminal part of the tomato 7S RNA and nucleotides 315-249 of PSTV. From the alignment it becomes evident that this complementarity is not continuous, but interrupted by short stretches of noncomplementary nucleotides. In the canonical rod-shaped secondary structure model of PSTV the site of complementarity is located in the left hand part of its lower strand (Fig. 3B). There is comprises most of the nucleotides of the premelting loop 1 region which is part of the previously defined (Schnölzer et al. 1985) virulence-modulating (VM) region of PSTV. The complementarity extends towards the right hand side and includes the lower strand of the central conserved (CC) region (Fig. 3A).

The sequence complementarity between tomato 7S RNA and PSTV (see Fig. 6) suggests that both RNA's could interact by forming a complex in vivo which might be of functional importance for the PSTV-infected tomato plant. Since this complementarity comprises a part of the VM region of PSTV, we assume that complex formation between both RNA could occur in vivo and could in some way be responsible for disease induction. The calculated free energy value of the complex between 7S RNA and the various PSTV isolates is about -45 kcal/mol. All these features provide the structural and thermodynamic preconditions that 7S RNA of tomato might be the cellular target for PSTV.

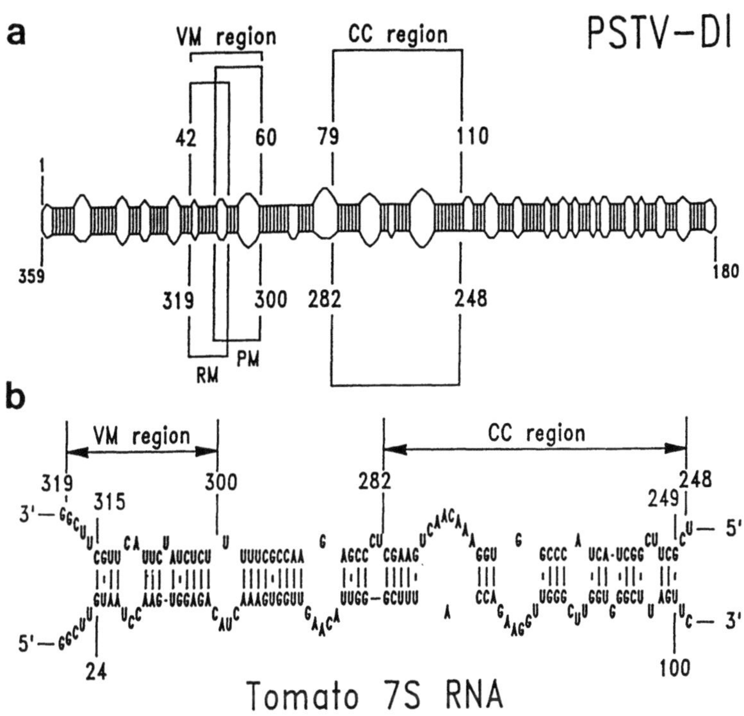

Fig. 6a,b. Site of possible base-pairing between PSTV and tomato 7S RNA. a Location of the central conserved (*CC*) region, and the virulence-modulating (*VM*) region with its region of mutation (*RM*) and the region of premelting loop 1 (*PM*). b Possible base-pairing scheme between the PSTV type strain (PSTV-DI) and tomato 7S RNA

Regarding the biological function of the plant 7S RNA, we assumed that it plays a similar role as the 7SL RNA of mamalian cells which is an integral part of the so-called signal recognition particle (SRP) (Li et al. 1982; Ullu et al. 1982; Walter and Blobel 1982). These SRP's have been demonstrated to be involved in the translocation of certain secretory and membrane proteins across the membrane of the endoplasmatic reticulum (ER) or the integration of these proteins into the membrane of the ER (Walter et al. 1984). Although the overall similarity between the sequence of the tomato 7S RNA and the sequence of the human 7SL RNA is not significant, both RNAs exhibit, nevertheless, very similar features in their secondary structure. This strongly suggests that also a functional similarity exists between the 7S RNA's from plant and mammalian cells. Recent investigations have shown that a SRP-like component is present in embryonic plant (wheat germ) tissue, that these particles also contain a 7S RNA, and that the wheat germ translocation system is similar to the mammalian one (Prehn et al. 1987).

Regarding the possible mechanism of viroid pathogenesis it is conceivable that due to the interaction between PSTV and the tomato 7S RNA, the normal function of this host RNA is impaired. If so, one would expect that viroid infections cause primarily cytopathological changes of the cell membranes. The interaction between the viroid and its cellular target could disturb the SRP-mediated translocation of membrane proteins and their integration into the endoplasmatic reticulum which would result in an abnormal formation of the cellular membranes. In fact, electron microscopical studies have shown that in viroid-infected plant tissue the plasma membranes, the membrane-associated plasmalemmasomes and the cell walls are malformed and extremely distorted (Wahn et al. 1980a,b; Momma and Takahaski 1983; Rosenberg et al. 1985). It has been found, in addition, that the sugar and amino acid composition of isolated cell walls from viroid-infected leaf tissue is markedly altered as compared with the one of noninfected tissue (Wang et al. 1986). Future work will have to show whether these alterations are based on the molecular interaction between viroids and the 7S RNA of their host or whether another mechanism is actually operating.

Concluding Remarks

The model we propose for the pathogenicity of PSTV in molecular terms is based on the primary and secondary structure and on thermodynamic properties of a distinct virulence-modulating region of the PSTV RNA. Our hypothesis postulates a competitive interaction between the PSTV RNA molecule and the 7S RNA as a specific cellular host factor(s). We would like to point out, however, that the proposed mechanism does not pretend to exclude other possible modes by which these unique pathogens might incide disease in higher plants.

Regarding our future work with viroids in general, our sequencing data of PSTV and CEV isolates provide the rationale for the direct analysis of viroid virulence by site-specific mutagenesis of viroids at the level of their cloned DNA. From this DNA a potentially infectious viroid RNA can be synthesized in vitro with the various RNA polymerase systems and then bioassayed on the appropriate testplants as already described for PSTV (Tabler and Sänger 1985).

In addition to this approach of utilizing recombinant DNA technology, the sequence analysis of many more naturally occurring viroids and viroid isolates should not be neglected. Such work might provide new insights into the possible mode of interaction between the viroid RNA and the putative host factors, in that other regions than the VM region could be found which contribute to the modulation of viroid virulence. In this way possible hints as to the nature of the cellular target could be obtained and the postulated role of the 7S RNA could be substantiated or refuted.

The experimental situation is much more complex if one intends to search for the host target at which viroid disease is incited. First of all we are faced with the fact that all the observed symptoms of viroid disease are the result of a cascade of cellular events which does not allow a reliable conclusion as to the interaction between the viroid RNA and its primary cellular target(s). In fact, plants are known to respond to infections with various microbial and viral pathogens in a similar way, although those may differ fundamentally in their their mode of interaction with the host plant.

To circumvent these system-inherent problems new experimental approaches are required. One promising way is the cross-linking of the complex between viroids and the cellular proteins or nucleic acids and the subsequent identification of these host components. But the question still remains whether or not such complexes are of functional relevance for viroid replication and pathogenicity.

Our finding of remarkable sequence complementarity between PSTV and tomato 7S RNA shows that the comparison of sequences may also lead to new concepts. Thus, the sequence analysis of the 7S RNA from other plant species will show if our hypothesis of 7S RNA as the primary host cell target for viroids can also be extended the various other viroid-host plant combinations. There is little doubt that the further elucidation of the mechanism of viroid replication and pathogenesis will also contribute to a better understanding of the normal processes of plant growth and development and their regulation.

References

Branch AD, Robertson HD, Dickson E (1981) Longer-than-unit-length viroid minus strands are present in RNA from infected plants, Proc Nat Acad Sci USA 78:6381-6385

Camacho Henriquez A, Sänger HL (1982a) Gelelectrophoretic analysis of phenol-extractable leaf proteins from different viroid/host combinations, Arch Virol 74: 167-180

Camacho Henriquez A, Sänger HL (1982b) Analysis of acid-extractable tomato leaf proteins after infection with a viroid, two viruses and a fungus and partial purification of the "pathogenesis-related" protein p14, Arch Virol 74:181-193

Camacho Henriquez A, Sänger HL (1984) Purification and partial characterization of the major "pathogenesis-related" tomato leaf protein P14 from potato spindle tuber viroid (PSTV)-infected tomato leaves. Arch Virol 81:263-284

Carey J, Cameron V, de Haseth PL, Uhlenbeck OC (1983) Sequence-specific interaction of R17 coat protein with its ribonucleic acid binding site. Biochemistry 22: 2601-2610

Conejero V, Picazo I, Segado P (1979) Citrus exocortis viroid (CEV):protein alterations in different hosts following viroid infection, Virology 97:454-456

Da Graca JV, Van Fuuren SP (1980) Transmission of avocado sunblotch disease to cinnamon. Plant Dis 64:475

Davies JW, Kaesberg P, Diener TO (1974) Potato spindle tuber viroid. XII. An investigation of viroid RNA as a messenger for protein synthesis, Virology 61:281-286

Dickson E (1979) Viroids: infectious RNA in plants. In: Nucleic Acids in Plants, Vol. 2, ed. T.C. Hall and J.W. Davies, pp 153-193

Dickson E, Robertson HD, Niblett CL, Horst RK, Zaitlin M (1979) Minor differences betweeen nucleotide sequences of mild and severe strains of potato spindle tuber viroid. Nature 277:60-62

Dickson E (1981) A model for the involvement of viroids in RNA splicing. Virology 115:216-221

Diener TO (1971) Potato spindle tuber virus: A plant virus with properties of a free nucleic acid. III. Subcellular location of PSTV-RNA and the question of whether virions exist in extracts or in situ, Virology 43:75-89

Diener TO (1977) Viroids: Autoinducing regulatory RNAs? In: Genetic interaction and gene transfer (ed. C.W. Anderson) (Brookhaven Symp. Biol.) Vol. 29, pp 50-61

Diener TO (1978) Are viroids autoinducing regulatory RNAs? In: Persistent Viruses, ed. J.G. Stevens, G.J. Todaro and C.F. Fox, pp 297-309. New York: Academic Press

Diener TO (1979) Viroids and viroid diseases. Wiley and Sons, New York

Diener TO (1981a) Are viroids escaped introns? Proc Natl Acad Sci USA 78:5014-5015

Dienter TO (1981b) Viroid: Abnormal products of plant metabolism, Ann Rev Plant Physiol 32:313-325

Diener TO (1983) Viroids, Advances in Virus Research 28:241-283

Duran-Vila N, Semancik JS (1982) Effects of exogenous auxins on tomato tissue infected with the citrus exocortis viroid. Phytopathology 72:777-781

Fernow KH (1967) Tomato as a test plant for detecting mild strains of potato spindle tuber virus. Phytopathology 57:1347-1352

Flores R, Chroboczek J, Semancik, JS (1978) Some properties of the CEV-P_1 protein from citrus exocortis viroid-infected Gynura aurantiaca D.C. Physiol Plant Pathol 13:193-201

Flores R (1984) Is the conformation of viroids involved in their pathogenicity? J theor Biol 108:519-527
Galindo AJ, Smith DR, Diener TO (1984) A disease-associated host protein in viroid-infected tomato. Physiol Plant Pathol 24:257-275
Garret RA, Douthwaite S, Noller HF (1981) Structure and role of 5S RNA-protein complexes in protein biosynthesis. Trends Biochem Sci 6:137-139
Goodman TC, Nagel L, Rappold W, Klotz G, Riesner D (1984) Viroid replication: equilibrium association constant and comparative activity measurements for the viroid-polymerase interaction. Nucleic Acid Res 12:6231-6246
Gralla J, Steits JA, Crothers DM (1974) Direct physical evidence for secondary structure in an isolated fragment of R17 bacteriophage mRNA. Nature 248:204-208
Gross HJ, Domdey H, Lossow C, Jank P, Raba M, Alberty H, Sänger HL (1978) Nucleotide sequence and secondary structure of potato spindle tuber viroid. Nature 273:203-208
Gross HJ, Krupp G, Domdey H, Steger G, Riesner D, Sänger HL (1981) The structure of three plant viroids. Nucleic Acids Res Symposium Series 10:91-98
Gross HJ, Krupp G, Domdey H, Raba M, Alberty H, Lossow CH, Ramm K, Sänger HL (1982) Nucleotide sequence and secondary structure of citrus exocortis and chrysanthemum stunt viroid. Europ J Biochem 121:249-257
Hall TC, Wepprich RK, Davies JW, Weathers LG, Semancik JS (1974) Functional distinctions between the ribonucleic acids from citrus exocortis viroid and plant viruses: Cell-free translation and aminoacylation reactions. Virology 61:486-492
Haseloff J, Symons RH (1981) Chrysanthemum stund viroid-primary sequence and secondary structure. Nucl Acids Res 9:2741-2752
Haseloff J, Mohamed NA, Symons RH (1982) Viroid RNAs of cadang cadang disease of coconuts. Nature 299:316-321
Imperial JS, Bautista RM, Randles JW (1985) Transmission of the coconut cadang cadang viroid to six species of palm by inoculation with nucleic acid extracts. Plant Pathol 34:391-401
Jakab G, Kiss T, Solomosy F (1986) Viroid pathogenicity and pre-rRNA processing: a model amenable to experimental testing. Biochim Biophys Acta 868:190-197
Kiefer MC, Owens RA, Dienter TO (1983) Structural similarities between viroids and transposable genetic elements. Proc Natl Acad Sci USA 80:6234-6238
Krug M, de Haseth PL, Uhlenbeck OC (1982) Enzymatic synthesis of a 21-nucleotide coat protein binding fragment of R17 ribonucleic acid. Biochemistry 21:4713-4720
Li WY, Reddy R, Henning D, Epstein P, Busch HJ (1982) Nucleotide sequence of 7S RNA: Homology to Alu DNA and La 4.5S RNA. J Biol Chem 257:5136-5142
Momma T, Takahashi T (1983) Cytopathology of shoot apical meristem of hop plants infected with hop stunt viroid. Phytopath Z 106:272-280
Mühlbach H-P, Sänger HL (1977) Multiplication of cucumber pale fruit viroid in inoculated tomato leaf protoplasts. J Gen Virol 35:377-386
Mühlbach H-P, Camacho-Henriquez A, Sänger HL (1977) Isolation and properties of protoplasts from leaves of healthy and viroid infected tomato plants. Plant Science Letters 8:183-189
Mühlbach H-P, Sänger HL (1981) Continous replication of potato spindle tuber viroid (PSTV) in permanent cell-cultures of potato and tomato. Biosci Rep 1:79-87
Mühlbach H-P, Faustmann O, Sänger HL (1983) Conditions for optimal growth of a PSTV-infected potato cell suspension and detection of viroid-complementary longer-than-unit-length RNA in these cells. Plant Molec Biol 2:239-247
Ohno T, Takamatsu N, Meshi T, Okada Y (1983) Hop stunt viroid: molecular cloning and nucleotide sequence of the complete cDNA copy. Nucleic Acids Res 11:6185-6197
Prehn S, Wiedmann M, Rapoport TA, Zwieb C (1987) Protein translocation acros wheat germ microsomal membranes requires an SRP-like component. EMBO J 6:2093-2097
Rackwitz HR, Rohde W, Sänger HL (1981) DNA-dependent RNA polymerase II of plant origin transcribes viroid RNA into full-length copies. Nature 291:297-301
Randles JW, Rillo EP, Diener TO (1976) The viroid-like structure and cellular location of anomalous RNA associated with the cadang-cadang disease. Virology 74:128-139
Randles JW, Boccardo G, Imperial JS (1980) Detection of cadang-cadang associated RNA in African oil palm and buri palm. Phytopathology 70:185-189

Reanney CD (1975) A regulatory role for viral RNA in eucaryotes. Journal of Theoretical Biology 49:461-492
Riesner D, Henco K, Rokohl U, Klotz G, Kleinschmidt AK, Gross HJ, Domdey H, Sänger HL (1979) Structure and structure formation of viroids. J Mol Biol 133:85-115
Riesner D, Steger G, Schumacher J, Gross HJ, Sänger HL (1981) Structure and function of viroids. Biophys Struct Mech 7:240-241
Riesner D, Steger G, Schumacher J, Gross HJ, Randles JW, Sänger HL (1983) Structure and function of viroids. Biophys Struct Mechanics 9:145-170
Riesner D, Gross HJ (1985) Viroids. Ann Rev Biochem 54:531-564
Robertson HD, Dickson E (1974) RNA processing and the control of gene expression. In: Processing of RNA. Brookhaven Symp Biol 26:240-266
Rodriguez JL, Garcia-Martinez JL, Flores R (1978) The relationship between plant growth substance content and infection of Gynura aurantiaca D.C. by citrus exocortis viroid. Physiol Plant Pathol 13:355-363
Rohde W, Sänger HL (1981) Detection of complementary RNA intermediates of viroid replication by northern blot hybridization. Biosci Rep 1:327-336
Rosenberg F, Wahn K, Sänger HL (1985) Cytopathologie von viroid-infiziertem Pflanzengewebe. II. Licht- und elektronenmikroskopische Untersuchungen am Blattgewebe von Chrysanthemum morifolium der Sorte "Mistletoe" nach Infektion mit dem Viroid der Stauchekrankheit der Chrysantheme (CSV). Phytopath Z 114:41-68
Sänger HL, Ramm K (1975) Radioactive labelling of viroid-RNA. In: "Modifications of the information content of plant cells," (R. Markham, D.R. Davies, D.A. Hopwood, and R.W. Horne, eds.), pp 229-252, North-Holland Publ Comp, Amsterdam
Sänger HL, Klotz G, Riesner D, Gross HJ, Kleinschmidt AK (1976) Viroids are single-stranded covalently closed circular RNA molecules existing as highly basepaired rod-like structures. Proc Natl Acad Sci USA 73:3852-3856
Sänger HL (1982) Biology, structure, functions and possible origin of viroids. In: "Nucleic Acids and Proteins in Plants II. Encyclopedia of Plant Physiology. New Series, Vol. 14B," (B. Parthier and D. Boulter, eds.), pp. 368-454, Springer-Verlag, Berlin, Heidelberg
Sänger HL (1984) Minimal infectious agents: The viroids. In: "The Microbe, Part I, Viruses" (B.W.J. Mahy and J.R. Pattison, eds.), pp 281-334, Society for General Microbiology Ltd., Cambridge University Press
Sano T, Uyeda I, Shikata E, Ohno T, Okada Y (1984) Nucleotide sequence of cucumber pale fruit viroid: homology to hop stunt viroid. Nucleic Acids Res 12:3427-3434
Schnölzer M, Haas B, Ramm K, Hofmann H, Sänger HL (1985) Correlation between structure and pathogenicity of potato spindle tuber viroid (PSTV), EMBO J 4:2181-2190
Schumacher J, Randles JW, Riesner D (1983a) A two-dimensional electrophoretic technique for the detection of circular viroids and virusoids. Analytical Biochemistry 135:288-295
Schumacher J, Sänger HL, Riesner D (1983b) Subcellular localization of viroids in highly purified nuclei from tomato leaf tissue. The EMBO J 2:549-1555
Semancik JS, Weathers LG (1972) Exocortis disease: evidence for a new species of "infectious" low molecular weight RNA in plants. Nature 237:242-246
Semancik JS, Conjero V, Gerhart J (1977) Citrus exocortis viroid: Survey of protein synthesis in Xenopus laevis oocytes following addition of viroid RNA. Virology 80: 218-221
Semancik JS (1979) Small pathogenic RNA in plants - the viroids. Ann Rev Phytopathol 17:461-484
Semancik JS, Rivera-Bustamante R, Goheen AC (1987) Widespread occurrence of viroid-like RNAs in grapevines. Am J Enol Vitic 38:35-40
Spiesmacher E, Mühlbach H-P, Schnölzer M, Haas B, Sänger HL (1983) Oligomeric forms of potato spindle tuber viroid (PSTV) and of its complementary RNA are present in nuclei isolated from viroid-infected potato cells. Biosci Rep 3:767-774
Spiesmacher E, Mühlbach H-P, Tabler M, Sänger HL (1985) Synthesis of (+) and (-)RNA molecules of potato spindle tuber viroid (PSTV) in isolated nuclei and its impairment by transcription inhibitors. Biosci Rep 5:251-265
Steger G, Hofmann H, Förtsch J, Gross HJ, Randles JW, Sänger HL, Riesner D (1984) Conformational transitions in viroids and virusoids: Comparison of results from energy minimization algorithm and from experimental data. J Biomol Struct Dyn 2:543-571

Symons RH (1981) Avocado sunblotch viroid: primary sequence and proposed secondary structure. Nucleic Acids Research 9:6527-6537

Tabler M, Sänger HL (1984) Cloned single- and double-stranded DNA copies of potato spindle tuber viroid (PSTV) and co-inoculated subgenomic fragments are infectious. EMBO J:30-5-3062

Tabler M, Sänger HL (1985) Infectivity studies on different potato spindle tuber viroid (PSTV) RNAs synthesized in vitro with the SP6 transcription system. EMBO J 4:2191-2199

Tabler M, Schnölzer M, Sänger HL (1985) Molecular cloning of potato spindle tuber viroid (PSTV) cDNA synthesized by enzymatic elongation of PSTV-specific DNA primers: A general strategy for viroid cloning. Biosci Rep 5:143-158

Tsagris M, Tabler M, Mühlbach H-P, Sänger HL (1987) Linear oligomeric potato spindle tuber viroid (PSTV) RNAs are accurately processed in vitro to the monomeric circular viroid proper when incubated with a nuclear extract from healthy potato cells. EMBO J 6:2173-2183

Ullu E, Murphy S, Melli M (1982) Human 7SL RNA consits of a 140 nucleotide middle-repetitive sequence inserted in an Alu sequence. Cell 29:195-201

Visvader JE, Gould AR, Bruening GE, Symons RH (1982) Citrus exocortis viroid: nucleotide sequence and secondary structure of an Australian isolate. FEBS Lett 137:288-292

Visvader JE, Symons RH (1983) Comparative sequence and structure of different isolates of citrus exocortis viroid. Virology 130:232-236

Visvader JE, Symons RH (1985) Eleven new sequence variants of citrus exocortis viroid and the correlation of sequence with pathogenicity. Nucleic Acids Res 13: 2907-2920

Visvader JE, Forster AC, Symons RH (1985) Infectivity and in vitro mutagenesis of monomeric cDNA clones of citrus exocortis viroid indicates the site of processing of viroid precursors. Nucleic Acids Research 13:5843-5856

Visvader JE, Symons RH (1986) Replication of in vitro constructed viroid mutants: location of the pathogenicity-modulating domain of citrus exocortis viroid. EMBO J 5:2051-2055

Wahn K, Rosenberg de Gomez F, Sänger HL (1980a) Cytopathologie von viroid-infiziertem Pflanzengewebe. I. Veränderungen des Plasmalemmas und der Zellwand bei Gynura aurantiaca DC nach Infection mit dem Viroid der Citrus Exocortis Krankheit (CEV). Phytopath Z 98:1-18

Wahn K, Rosenberg de Gomez F, Sänger HL (1980b) Cytopathic changes in leaf tissue of Gynura aurantiaca infected with the viroid of the Exocortis disease of citrus. J gen Virol 49:355-365

Walter P, Blobel G (1982) Signal recognition particle contains a 7S RNA essential for protein translocation across the endoplasmic reticulum. Nature 299:691-698

Walter P, Gilmore R, Blobel G (1984) Protein translocation across the endoplasmic reticulum. Cell 38:5-8

Wang MC, Lin JJ, Duran-Vila N, Semancik JS (1986) Alteration in cell wall composition and structure in viroid-infected cells. Physiol Molec Plant Pathol 28:107-124

Wolff P, Gilz R, Schumacher J, Riesner D (1985) Complexes of viroids with histones and other proteins. Nucleic Acids Res 13:355-367

Zaitlin M, Hariharasubramanian V (1972) A gel electrophoretic analysis of proteins from plants infected with tobacco mosaic and potato tuber viruses. Virology 47: 296-305

Zimmern D (1982) Do viroids and RNA viruses derive from a system that exchanges genetic information between eukaryotic cells? Trends in Biochemical Science 8:205-207

Antigenicity of the Influenza Haemagglutinin Membrane Glycoprotein

J. J. Skehel[1] and D. C. Wiley[2]

Antibodies which neutralize the infectivity of influenza viruses specifically react with the hemagglutinin membrane glycoproteins. These molecules, of which there are about 500 on each virus surface, serve to bind viruses to sialic acid-containing cellular receptors and to mediate fusion of virus and cellular membranes required for entry of virus genome-transcriptase complexes into the cell (reviewed in [1]). Neutralization probably involves inhibition of these functions. Hemagglutinins (HA) are trimers of molecular weight about 220,000 in which each monomer consists of two disulfide-linked glycopolypeptides, HA_1 and HA_2. The three-dimensional structure of the Hong Kong influenza X-31 HA has been determined crystallographically to 3Å resolution [2]. The trimer is 135 Å long and approximately triangular in cross-section, varying in radius from about 15 to 40 Å. In each subunit the HA_1 glycopolypeptide chain extends from the base of the molecule near the virus membrane, through a fibrous stem to a peripheral region rich in β-structure, and then returns to the fibrous region and terminates about 30 Å from the membrane. The most prominent features of the part of the subunit composed of HA_2 residues are two antiparallel α-helices, one 29 Å long which proceeds distally to connect through an extended chain with the other helix, which stretches 76 Å back towards the membrane. This paper briefly describes studies of the antigenic variation of this molecule and its recognition by both antibodies and immune T cells.

The progressive change in the antigenic properties of HA which occurs with time after the pandemics which mark the introduction of a new subtype of influenza A into the human population is called antigenic drift. The changes in antigenicity are readily analyzed in hemagglutination inhibition tests using convalescent ferret antisera, and recently these analyses have been accompanied by studies of the nucleotide sequences of the RNA genes for HA of representative isolates [3,4,5,6,7]. The results obtained indicate that antigenic drift results from the accumulation of amino acid substitutions in the HA_1 glycopolypeptide of the HA subunit. In the H_3 subtype, since 1968 73 of the 328 residues of HA_1 have been observed to have changed compared with only 12 in HA_2 [3,4,8]. The substitutions in HA_1 cluster in five surface regions of the distal domain labelled A to E [9]. They involve changes in charge, e.g., HA_1 144 gly to asp, HA_1 156 lys to glu, or side-chain length, e.g., HA_1 198 ala to val, or changes which lead to additional sites for glycosylation: HA_1 63 asp to asn in the sequence asn-cys-thr, HA_1 124 gly to ser in the sequence asn-glu-ser, and HA_1 248 asn to ser or thr in the sequence asn-ser-/thr. For the first of the three last examples a direct effect of the carbohydrate side chain on HA antigenicity has been shown [10] and the other two extra sites for glycosylation are in positions where a carbohydrate side chain could mask antigenically important regions of the molecule. In some positions, e.g., HA_1 144 gly to asp to val, repeated changes occur.

---

[1] National Institute for Medical Research, The Ridgeway, Mill Hill, London NW7 1AA, UK

[2] Department of Biochemistry & Molecular Biology, Harvard University, Cambridge, Massachusetts 02138, USA

38. Colloquium Mosbach 1987
Molecular Basis of Viral and Microbial Pathogenesis

The antigenic significance of changes in these five regions is confirmed by observations that single amino acid substitutions in one or other of the sites are detected in the HA's of monoclonal antibody selected antigenic variants [11,12,13,14,15]. Two such variants containing HA_1 146 gly to asp or HA_1 188 asn to asp substitutions have also been used to provide evidence that antibodies bind to the region of the molecule in which the amino acid substitution occurs [16]. In these cases the structures of the mutant HA's were determined crystallographically and the structural differences from the wild-type HA shown to be confined to the immediate vicinity of the amino acid substitution. Such observations also indicate directly that the simple addition of an amino acid side chain is a sufficient structural alteration to prevent antibody binding. The results of electron microscopy of HA-monoclonal antibody complexes for antibodies which recognize changes in sites A, B and E are consistent with the conclusion that the site of amino acid substitution is within the site of antibody binding [17].

Overall, the information obtained from these studies of antigenic drift indicates that as a result of immune pressure antigenic variants are selected which contain amino acid substitutions at the surface of the membrane distal domain of the molecule. These substitutions prevent the binding of antibodies induced following a previous infection and, as a consequence, the variants selected have the ability to re-infect. The direction of antigenic drift is dictated by specific amino acid substitutions which are shared by subsequent variants, and viruses which cause major epidemics contain substitutions in each of the antigenically important regions.

Considering that the frequency at which antigenic variants are selected by monoclonal antibodies is between 10^{-4} and 10^{-5} [18], and that antibodies which recognize any of the five antigenic regions neutralize virus infectivity, selection of a mutant with the ability to re-infect the majority of the population would be expected to occur spontaneously at a very low frequency. Changes at each of the antigenic sites may, on the other hand, occur during re-infections by antigenic mutants changed in only one or two sites of individuals who develop only partial immunity during initial infections. Analysis of the variety of antibody specificities in post-infection human sera [19] and of the restricted ability of human sera to neutralize monoclonal antibody selected mutants [20] suggest that this is the case.

Immune T cells differ from antibodies which bind to native HA on viruses to neutralize infectivity, in recognizing fragments of the HA in association with major histocompatibility antigens. HA recognition by helper T cells appears to involve primarily the HA_1 glycopolypeptide including a number of regions that are recognized by antibodies. HA-specific helper T-cell lines derived from peripheral blood lymphocytes from humans were found to be stimulated to proliferate by peptides HA_1 1-36, 105-140, 200-228, 306-328 of H_3 subtype HA and to a lesser extent by synthetic peptides approximately 20 amino acids in length, equivalent to all regions of HA_1. The most immuno-dominant peptide, HA_1 306-323 was recognized by both subtype-specific and cross-reactive clones isolated from these T cell lines [21,22]. Helper T cells induced by infection of mice with A/PR/8/34 virus of the H_1 subtype were found to recognize three regions of the HA, two of which were defined using peptides HA_1 109-120 and HA_1 290-310. The recognition specificity of clones which responded to the former peptide was analyzed using fragments of HA_1, various synthetic peptides from the HA_1 109-120 region and a series of natural and monoclonal antibody selected variants. As a result, the importance of HA_1 115 for recognition was established [23,24,25,26]. The third epitope identified in these studies included

HA_1 136. H_3 subtype HA-specific mouse helper T cell clones have also been isolated which recognize synthetic peptide analogues of HA_1 48-68 and HA_1 118-138 and the range of specificities of as yet undefined clones suggests the importance of additional regions of HA_1 in recognition [27,28]. Furthermore, in other studies in which synthetic peptides and isolated HA_1 and HA_2 glycopolypeptides were used to stimulate proliferation, helper T cells that recognize components of HA_2 were also detected, and evidence was presented that they were more subtype cross-reactive than cells which recognize HA_1 [29].

Finally, cytotoxic T cells also appear to recognize both HA_1 and HA_2 components of HA. Induction of secondary cytotoxic cells was observed using a cyanogen bromide fragment, HA_2 103-123 and confirmed by using an equivalent synthetic peptide. Also, a peptide analog of HA_1 181-204 in the same study was shown to induce at about ten times greater molar efficiency. Both peptides were recognized by H_2 subtype-specific cytotoxic cells [30,31]. Experiments with a fusion protein containing HA_2 produced in *E. coli* have also shown that this region of the haemagglutinin can generate target cells recognized by subtype-specific cytotoxic cells [32]. Understanding the role of these cells in immunity to influenza is an objective of current research.

References

1. Wiley DC, Skehel JJ (1987) Annu Rev Biochem in press
2. Wilson IA, Skehel JJ, Wiley DC (1981) Nature (London) 289:368-373
3. Both GW, Sleigh MJ, Cox NJ, Kendal AP (1983) J Virol 48:52-60
4. Skehel JJ, Daniels RS, Douglas AR, Wiley DC (1983) Bull WHO 61:671-676
5. Daniels RS, Douglas AR, Skehel JJ, Wiley DC (1985) Bull WHO 63:273-277
6. Raymond FL, Caton AJ, Cox NJ, Kendal AP, Brownlee GG (1983) Nucleic Acids Res 11:7191-7203
7. Raymond FL, Caton AJ, Cox NJ, Kendal AP, Brownlee GG (1986) Virology 148:275-287
8. Stevens DJ, Skehel JJ, Wiley DC (1987) Bull WHO in press
9. Wiley DC, Wilson IA, Skehel JJ (1981) Nature (London) 289:373-378
10. Skehel JJ, Stevens DJ, Daniels RS, Douglas AR, Knossow M, Wilson IA, Wiley DC (1984) Proc Natl Acad Sci USA 81:1779-1783
11. Laver WG, Air GM, Webster RG, Gerhard W, Ward CW, Dopheide TA (1979) Virology 98:226-237
12. Laver WG, Air GM, Webster RG (1981) J Mol Biol 145:339-361
13. Daniels RS, Douglas AR, Skehel JJ, Wiley DC (1983) J Gen Virol 64:1657-1661
14. Newton SE, Air GM, Webster RG, Laver WG (1983) Virology 128:495-501
15. Webster RG, Brown LE, Jackson DC (1983) Virology 126:587-599
16. Knossow M, Daniels RS, Douglas AR, Skehel JJ, Wiley DC (1984) Nature (London) 311:678-680
17. Wrigley NG, Brown EB, Daniels RS, Douglas AR, Skehel JJ, Wiley DC (1983) Virology 131:308-314
18. Yewdell JW, Webster RG, Gerhard W (1979) Nature (London) 279:246-248
19. Wang M-L, Skehel JJ, Wiley DC (1986) J Virol 57:124-128
20. Natali A, Oxford J, Schild GC (1981) J Hyg 87:185-190
21. Lamb JR, Eckels DD, Lake P, Woody JN, Green N (1982) Nature (London) 300:66-69
22. Lamb JR, Green N (1983) Immunology 50:659-666
23. Hackett CJ, Dietzschold B, Gerhard W, Christ B, Knorr R (1983) J Exp Med 158: 294-302
24. Hurwitz JL, Haber-Katz E, Hacket CJ, Gerhard W (1984) J Immunol 133:3371-3377
25. Gerhard W, Hackett C, Melchers FJ (1983) J Immunol 130:2379-2385
26. Hacket CJ, Hurwitz JL, Dietzchold B, Gerhard W (1985) Immunology 135:1391-1394
27. Mills KHG, Skehel JJ, Thomas DB (1986) J Exp Med 163:1477-1490
28. Mills KHG, Skehel JJ, Thomas DB (1986) Eur J Immunol 16:276-280
29. Katz JM, Laver WG, White DO, Anders DM (1985) J Immunol 134:616-622

30. Wabuke-Bunoti MAN, Fan DP (1983) J Immunol 130:2386-2391
31. Wabuke-Bunoti MAN, Taku A, Fan DP, Kent S, Webster RG (1984) J Immunol 133:2194-2201
32. Yamada A, Young JF, Ennis FA (1985) J Exp Med 162:1720-1725

Viral Glycoproteins as Determinants of Pathogenicity

H.-D. Klenk[1], M. Tashiro[2], W. Garten[1], and R. Rott[2]

Introduction

Many viruses possess a lipid-containing envelope which is a simple biological membrane surrounding the viral genome and accessory proteins. Among the general features of membrane structure shared by all of these viruses are glycoproteins that span the lipid bilayer and usually appear as protrusions or spikes at the surface of the virus particles. Viral glycoproteins have found wide interest as models for structure and biosynthesis of membrane proteins. Moreover, they are the major targets for the humoral and cellular immune response against the virus. This chapter will deal with another important aspect of these glycoproteins, namely their key role in virus entry into the host cell and their significance as determinants of virus pathogenicity. Our knowledge on the role of viral glycoproteins as initiatiors of infection and determinants of pathogenicity has been derived mainly from studies on ortho-and paramyxoviruses. We will therefore concentrate here on these viruses, but we will also present evidence that similar concepts may also be relevant for other viruses.

Influenza Virus: Agents with a Broad Range of Pathogenicity

Influenza viruses have been isolated from man, horse, swine, mink, seal, and whale, as well as from a great variety of different avian species. The infection in mammals is usually confined to the respiratory tract, while in birds there is a systemic infection which in general leads to death. The majority of the avian influenza viruses, however, induce an assymptomatic local infection that is confined to the mucosa of the gut or the respiratory tract. Among the avian influenza viruses, 13 different hemagglutinin subtypes can be discriminated that occur in many hemagglutinin-neuraminidase combinations. The highly pathogenic strains contain only H7 and H5 hemagglutinin subtypes. As far as is known, the strains containing all other subtypes are apathogenic. Thus, the avian system is particularly useful for pathogenicity studies, because a large number of naturally occurring virus strains can be analyzed in their natural host.

The Influenza Virus Hemagglutinin Initiates Infection

Influenza A and B viruses contain two glycoproteins, hemagglutinin and neuraminidase. Hemagglutinin is one of the best-characterized membrane proteins. The amino acid sequence of several subtypes has been elucidated (for literature see Lamb 1983), and the tertiary structure has

[1]Institut für Virology, Phillips-Universität, Marburg, FRG

[2]Institut für Virology, Justus-Liebig-Universität, Gießen, FRG

38. Colloquium Mosbach 1987
Molecular Basis of Viral and Microbial Pathogenesis

been determined by X-ray crystallography (Wilson et al. 1981). It was shown that the hemagglutinin is a trimer containing three monomers that are composed of the polypeptide chains HA_1 and HA_2. HA_1 forms a globular domain located at the distal end of a stalk consisting mainly of HA_2. The hemagglutinin carries several N-glycosidic oligosaccharide side chains that exhibit site-specific variations in structure (Keil et al. 1985). A sequence of 25-32 hydrophobic amino acids near the C-terminus of HA_2 serves to anchor the hemagglutinin in the virus membrane.

As an integral membrane protein, the hemagglutinin is translated at membrane-bound polysomes, translocated by means of an amino-terminal signal sequence into the lumen of the endoplasmic reticulum and transported from there through the Golgi apparatus to the plasma membrane. In the course of transport, the hemagglutinin undergoes a series of co- and post-translational modifications. These include, in the rough endoplasmic reticulum, the removal of the signal sequence and the attachment of the sugar side chains and, in the Golgi apparatus, remodeling of the oligosaccharides and proteolytic cleavage of the precursor HA into the amino-terminal fragment HA_1 and the carboxy-terminal fragment HA_2 (Fig. 1). Cleavage of HA is exerted by cellular proteases (for references see Klenk and Rott 1980).

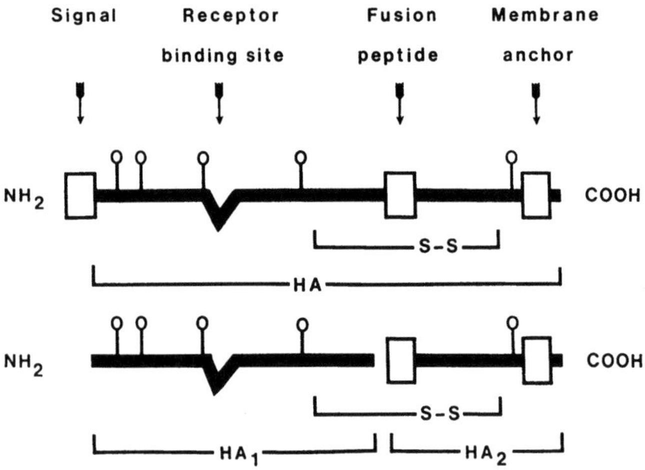

Fig. 1. Proteolytic cleavage of the hemagglutinin of influenza virus. The uncleaved precursor HA and the cleavage products HA_1 and HA_2 are shown. Hydrophobic sequences forming the signal, the fusion peptide, and the membrane anchor are boxed. The receptor binding site and oligosaccharide side chains are also indicated

Hemagglutinin plays an essential role in the initiation of infection, and it is now clear that it has two different functions in this process, by mediating adsorption of the virus particle to the cell surface as well as penetration of the viral genome into the cytoplasm. According to the available evidence the following steps can be discriminated: First, the hemagglutinin attaches by means of its receptor binding site on HA_1 to the neuraminic acid-containing receptor of the host cell (adsorption). Subsequently, a conformational change triggered by a shift to acidic pH (Skehel et al. 1982) may expose a hydrophobic region at the amino terminus of HA_2, the so-called fusion peptide, which can now insert into the target membrane, thereby forming a bridge between the two membranes. Alternatively, exposure of the hydrophobic amino terminus of HA_2 may result in an aggregation of the hemagglutinin in the plane of the membrane and in the concomitant formation of areas of the viral envelope devoid of protein. Both mechanisms could result in the close apposition of the lipid layers of the cellular and the viral membrane, and thus facilitate their fusion. Since the amino terminus of HA_2 is not exposed on the precursor HA, the fusion capacity of the hemagglutinin is activated only after proteolytic cleavage (Huang et al. 1981). Cleavage is therefore also a precondition for the infectivity of the virus (Klenk et al. 1975, Lazarowitz and Choppin 1975).

Although neuraminic acid is the essential receptor constituent for all influenza viruses, there are distinct variations in receptor specificity among different influenza viruses that are determined by the structure of the asialo-oligosaccharide, the linkage between the neuraminic acid and the oligosaccharide, and the substitution of the neuraminic acid (Rogers and Paulson 1983, Higa et al. 1985, Rogers et al. 1986). As will be outlined elsewhere in this volume, the receptor specificity of the hemagglutinin is an important determinant for tissue tropism and host range of influenza virus. It is likely that it contributes also to pathogenicity, but this concept has to be proven by further experimental data. Proteolytic activation of the fusion activity depends also on an intimate interplay between the hemagglutinin and cellular factors, and here it is quite clear that these interactions are important control mechanisms for pathogenicity.

Proteolytic Activation of the Influenza Hemagglutinin Determines Pathogenicity

Proteolytic activation of the hemagglutinin follows a pattern observed with many enzyme and hormone precursors, such as proinsulin, progastrin, proopiomelanocortin and others (Docherty and Steiner 1982). It involves cleavage at an arginine residue by the sequential action of a "trypsin-like" endoprotease and a carboxypeptidase that are both provided by the host.

Studies with specific inhibitors have revealed that, unlike the endoprotease, the carboxypeptidase is not necessary for activation (Garten and Klenk 1983). When a cell does not contain an appropriate endoprotease, virus with uncleaved hemagglutinin is released. Under these conditions, a whole series of other "trypsin-like" endoproteases, such as trypsin itself, plasmin (Lazarowitz et al. 1973), acrosin (Garten et al. 1981), or bacterial proteases (Klenk et al. 1977, Tashiro et al. 1987a,b) can substitute for the intracellular enzyme.

Since the activating proteases are cellular enzymes, the infected cell type determines whether or not the hemagglutinin is cleaved (Klenk et al. 1975, Klenk and Rott 1980). The hemagglutinins of the mammalian influenza viruses and the nonpathogenic avian influenza viruses, which cause a local infection, are susceptible to proteolytic cleavage only in a restricted number of cell types. On the other hand, the hemagglutinins of pathogenic avian viruses among the H5 and H7 subtypes, causing a systemic infection, are cleaved by proteases present in a broad range of different host cells (Bosch et al. 1979). Thus, there are differences in host range resulting from differences in hemagglutinin cleavability which can be correlated with the pathogenic properties of the virus. The differences in cleavability are due to differences in the structure of the cleavage site of the hemagglutinins. The HA_1 and HA_2 fragments of the hemagglutinin molecule of the apathogenic avian and of all mammalian influenza viruses are linked by a single arginine. This is in contrast to the pathogenic avian strains, which have a sequence of several basic amino acids at the cleavage site. This concept was first suggested by electrofocusing studies (Bosch et al. 1981, Garten et al. 1981), and has now been confirmed by sequence analyses

Table 1. Cleavage sites of the hemagglutinins of mammalian and avian influenza A viruses

Virus	Sero-type	Pathogenicity for fowl	Cleavage site	References
A/Puerto Rico/8/34	H1		-Pro-Ser-Ile-Gln-Tyr———ARG-∇Gly-Leu-	Winter et al. 1981
A/Japan/305/57	H2		-Pro-Gln-Ile-Gln-Ser———ARG-∇Gly-Leu-	Gething et al. 1980
A/Memphis/102/72	H3		-Pro-Glu-Lys-Gln-Thr———ARG-∇Gly-Leu-	Sleigh et al. 1980
A/Duck/Ukraine/1/63	H3	apathogenic	-Pro-Glu-Lys-Glu-Thr———ARG-∇Gly-Leu-	Fang et al. 1981
A/Chick/Penn/1370/83	H5	pathogenic	-Pro-Gln-LYS——LYS-LYS———ARG-▼Gly-Leu-	Kawaoka et al. 1984
A/Seal/Mass/1/80	H7		-Pro-Glu-Asn-Pro-Lys-Thr———ARG-∇Gly-Leu-	Naeve and Webster 1983
A/FPV/Rostock/34	H7	pathogenic	-Pro-Glu-Pro-Ser-Lys-Lys-Arg-Glu-LYS-ARG-▼Gly-Leu-	Porter et al. 1979
A/FPV/Rostock/34	H7	pathogenic	-Pro-Glu-Pro-Ser-LYS-LYS-ARG-LYS-LYS-ARG-▼Gly-Leu-	Garten et al. 1985
A/FPV/Dutch/27	H7	pathogenic	-Pro-Glu-Leu-Pro-LYS-LYS-ARG-ARG-LYS-ARG-▼Gly-Leu-	Pritzer 1985
A/Chick/Germany/49	H10	apathogenic	-Pro-Glu-Val-Val-Gln-Gly———ARG-∇Gly-Leu-	Feldmann 1988

∇ Cleavage site of endoprotease recognizing a single arginine.

▼ Cleavage site of endoprotease recognizing pairs of basic residues. Amino acids eliminated by the sequential action of endoprotease and carboxypeptidase are indicated by capitals.

(Table 1). Although the hemagglutinins of all influenza viruses are cleaved by the same general mechanism resulting in the elimination of the basic amino acids, it has to be assumed that differences exist in the specificity of the endoproteases that recognize either a single arginine or the paired basic residues lysine-arginine or arginine-arginine (Garten et al. 1982, Garten and Klenk 1983).

There is evidence that mutations in sequences different from the cleavage site can also affect the activation of the molecule. Such a mutation has been found to be responsible for an altered protease sensitivity, which in turn resulted in a change in host cell range (Rott et al. 1984).

A single point mutation in the hemagglutinin appears also to be responsible for the drastic increase in pathogenicity that occurred within a series of outbreaks of avian influenza in Pennsylvania in 1983. Comparison of different isolates which all belonged to the H5N2 subtype revealed that virus from an early outbreak had low pathogenicity and contained a hemagglutinin with restricted cleavability, whereas virus from a late outbreak was highly pathogenic and had a hemagglutinin that was always cleaved. Sequence analysis revealed that, as shown already in Table 1, the pathogenic variant had a cleavage site of the structure lysine-lysine-lysine-arginine, i.e., a cleavage site susceptible to ubiquitous proteases. The apathogenic precursor strain had the same cleavage site, but here it was masked by an adjacent oligosaccharide which was subsequently lost by point mutation (Kawaoka et al. 1984, Kawaoka and Webster 1985).

These data taken together indicate the important role of proteolytic activation of the hemagglutinin for pathogenicity. If the hemagglutinin is cleaved in a restricted number of cell types, the infection will be confined to localized areas of the host. In mammals, this type of infection affects the respiratory tract, whereas in birds it is likely to be clinically inapparent. On the other hand, cleavability of the hemagglutinin in a wider range of different host cells, as is the situation in the pathogenic H5 and H7 viruses, permits a rapid production of infectious virus particles in all organs and thereby spread in the organism, resulting in a systemic fatal disease (Rott et al. 1980, Klenk et al. 1984, Rott and Klenk 1986).

Cleavage Activity by Bacterial Proteases

There is now evidence that proteolytic activation of the hemagglutinin is a pathogenicity determinant not only for avian but also for mammalian influenza viruses. Combined viral-bacterial pneumonia in man is considered to be three times more common than primary viral pneumonia (Stuart-Harris et al.1985). Besides a number of different bacteria, *Staphylococcus aureus* is most commonly involved. It is generally thought that virus infection in a given tissue favours growth conditions for bacteria. On the other hand, it was recently found that Staphylococci exert a decisive effect on influenza virus replication in the respiratory tract and promote the development of influenza pneumonia. Some *Staphylococcus aureus* strains have been shown to secrete a protease capable of activating the hemagglutinin by proteolytic cleavage in vitro. The presence of the bacterial enzymes in cell culture media enabled the virus to undergo multiple growth cycles. Thus, co-infection of mice with *Staphylococcus aureus* enhanced the virus titer in the lung enormously, resulting in a fatal disease with extended lesions in lung tissue (Tashiro et al. 1987a,b) (Fig. 2). Development of influenza pneumonia could be prevented by treatment of infected mice with a protease

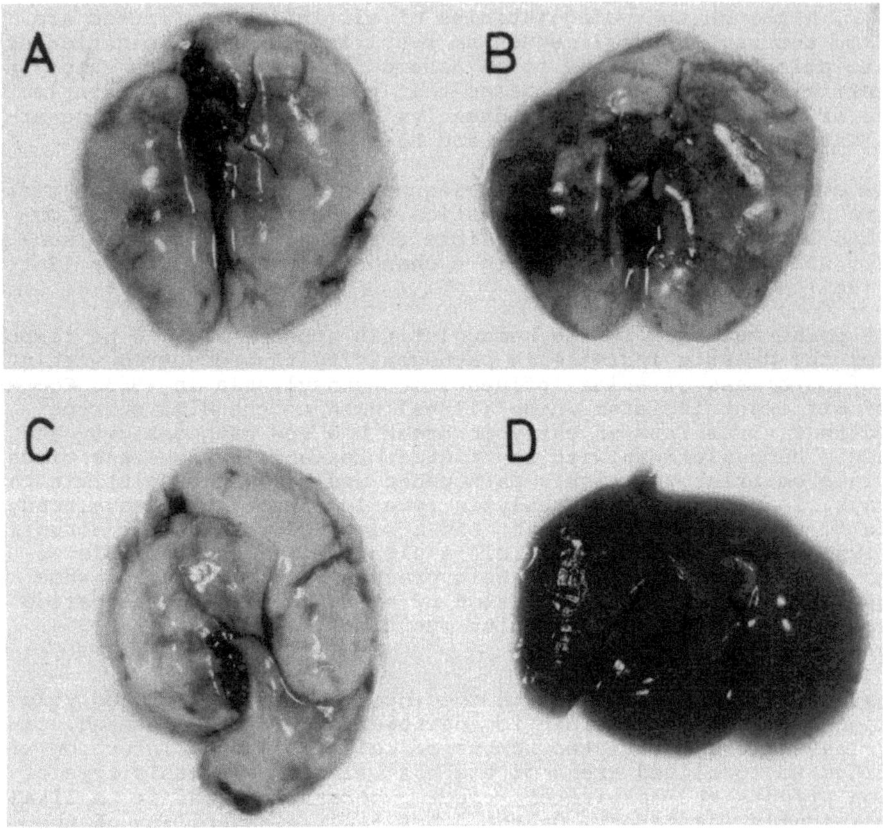

Fig. 2A-D. Pathological alterations in the lungs of mice after double infection with influenza virus and Staphylococcus aureus. *A* Mock-infected; *B* intranasally infected with A/swine/1976/31 (H1N1) virus; *C* infected with *Staphylococcus aureus* Wood 46; and *D* co-infected with virus and *Staphylococcus aureus*, respectively. The lungs were taken 5 days after infection. Doses of inoculum were 10^2 plaque-forming units of virus and 10^6 colony-forming units of bacterium. (Tashiro et al. 1987a)

inhibitor (Tashiro et al. 1987c). These findings may explain the high fatality rate seen after co-infection with *Staphylococcus aureus* in man (Robertson et al. 1958). *Staphylococcus aureus* is most likely not the only microorganism that can provide suitable proteases. One of several other candidates is *Haemophilus influenzae*, which was regularly isolated during the devastating influenza pandemics of 1918/19.

Cleavage Activation of the Glycoproteins of Other Viruses

Sequence analyses have revealed that post-translational proteolytic cleavage at arginine residues occurs also with the glycoproteins of many other viruses. These include the F protein of paramyxoviruses, the precursor to the E 2 protein of alphaviruses, the E 2 protein of coronaviruses, the GPC protein of arenaviruses, and the env protein of retroviruses (Tabel 2). In agreement with the observations made on the influenza virus hemagglutinin, glycoproteins with paired basic residues at the cleavage site are cleaved in all host cells. Virions containing

Table 2. Cleavage sites of different viral glycoproteins

Glycoprotein	Virus	Cleavage site	References
F-Protein of paramyxoviruses	Sendai virus	-Ala-Gly-Val-Pro-Gln-Ser-ARG- ▽ Phe-Phe	Blumberg et al. 1985
	SV5	-Pro-Thr-ARG-ARG-ARG-ARG-ARG- ▼ Phe-Ala-	Paterson et al. 1984
	RSV	-Ser-LYS-LYS-ARG-LYS-ARG-ARG- ▼ Phe-Leu-	Collins et al. 1984
	Measles Virus	-Ser-Ser-Arg-Arg-His-LYS-ARG- ▼ Phe-Ala-	Richardson et al. 1985
E2-Protein of alphaviruses	SFV	-Asn-Gly-Thr-Arg-His-ARG-ARG- ▼ Ser-Val-	Garoff et al. 1980
	Sindbis Virus	-Ser-Ser-Gly-Arg-Ser-LYS-ARG- ▼ Ser-Val-	Rice and Strauss 1981
E2-Protein of coronaviruses	IBV	-Gly-Thr-Arg-Arg-Phe-ARG-ARG- ▼ Ser-Thr-	Binns et al. 1985
Env-Protein of retroviruses	Rous sarcomavirus	-Thr-Gly-Ile-ARG-ARG-LYS-ARG- ▼ Ser-Val-	Schwarz et al. 1983
	HIV 1	-Pro-Thr-Lys-Ala-LYS-ARG-ARG- ▼ Val-Val-	Ratner et al. 1985
GPC of arenaviruses	Lassa virus	-Asp-Ile-Tyr-Ile-Ser-ARG-ARG- ▼ Leu-Leu-	Auperin et al. 1986

▽ Cleavage site of endoprotease recognizing a single arginine.

▼ Cleavage site of endoprotease recognizing pairs of basic residues.
Amino acids eliminated by the sequential action of endoprotease and carboxypeptidase are capitalized.

uncleaved precursor proteins can be obtained from appropriate cells only if the cleavage site consists of a single arginine.

This is the case with the F protein of Sendai virus, and it should be pointed out that this was the first viral glycoprotein shown to be activated by proteolytic cleavage (Homma and Ohuchi 1973, Scheid and Choppin 1974). The F protein mediates penetration of the paramyxovirus genome into the host cell by fusion of the viral membrane with the plasma membrane of the cell at neutral pH. Fusion, and thus virus infectivity depends on the cleavage of the precursor F_0 into the amino-terminal fragment F_2 and the carboxy-terminal fragment F_1 (Fig. 3). The new amino terminus of F_1 is hydrophobic and highly conserved among Sendai, SV5, and Newcastle disease virus (Scheid et al. 1978, Gething et al. 1978, Richardson et al. 1980), as well as among several other paramyxoviruses, including mumps (Server et al. 1985), measles (Richardson et al. 1986), and paramyxovirus 3 (Spriggs et al. 1986). Valuable information on the structural requirements determining the protease specificity of the F protein has been obtained from Sendai virus mutants of which the F protein could be activated by enzymes different from trypsin, such as elastase, chymotrypsin, or plasmin (Scheid and Choppin 1976). Sequence analysis revealed one or two amino acid changes near the cleavage site. In some cases, the change of protease specificity could be accounted for by changed amino acids right at the cleavage site, whereas several other mutations that potentiate cleavage at new sites by new proteases were somewhat removed from the actual cleavage site. It was assumed that such mutations might alter local peptide confirmation, thereby allowing the proteases access to existing sites (Hsu et al. 1987).

With most paramyxoviruses, adsorption is exerted by a second glycoprotein (HN) which has hemagglutinin and neuraminidase activity (Scheid et al. 1972). Unlike the F protein and the hemagglutinin of influenza viruses, HN is inserted with its amino-terminal end in the lipid bilayer (Schuy et al. 1984). With some strains of Newcastle disease virus HN is synthesized as a precursor HN_0 (Nagai et al. 1976). HN_0 is converted into HN by removal of a 9 kilodalton glycopeptide from its carboxy terminus (Garten et al. 1980) (Fig. 3). Cleavage is again a posttranslational event exerted by host proteases and necessary for full biological activity of the glycoprotein (Nagai and Klenk 1977).

Proteolytic Activation of the Glycoproteins Determines the Pathogenicity of Newcastle Disease Virus

Newcatle disease virus comprises a whole series of strains which, like the avian influenza viruses, differ widely in pathogenicity for chickens. Differences in pathogenicity can be correlated to the susceptibility of the envelope glycoproteins to proteolytic cleavage (Nagai et al. 1976), and recent sequence analyses have revealed that the apathogenic strains have F proteins with single arginine residues and the pathogenic strains F proteins with paired basic residues at their cleavage sites, exactly as has been observed with the avian influenza viruses (Toyoda et al. 1987) (Table 3). Mutants have been isolated from apathogenic strains, in which an enhanced susceptibility of the F proteins to cleavage is paralleled by an increase in pathogenicity (Garten et al. 1980).

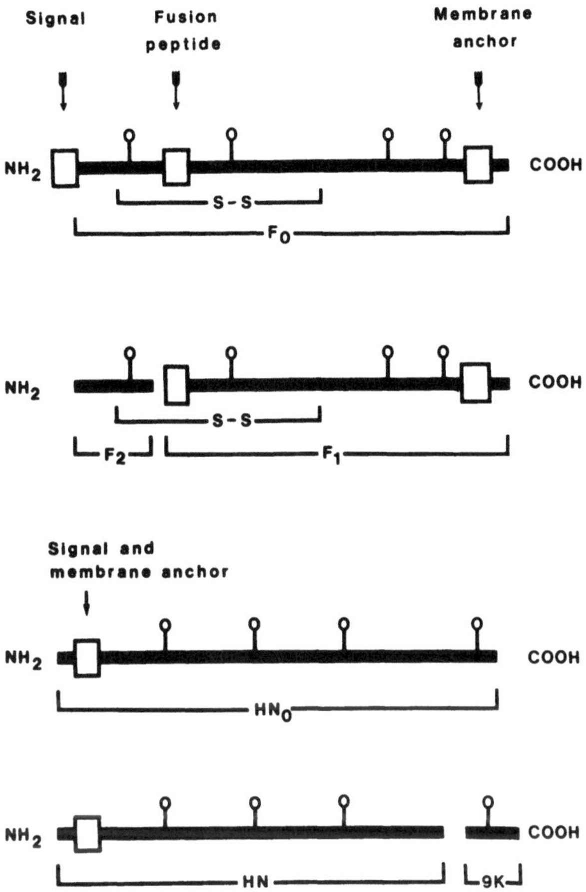

Fig. 3. Proteolytic cleavage of paramyxovirus glycoproteins. The precursor F_0 of the fusion protein is cleaved into the fragments F_1 and F_2 which are linked by disulfide bonds. Hydrophobic sequences forming the signal, the fusion peptide, and the membrane anchor are boxed. The precursor HN_0 of the hemagglutinin-neuraminidase is cleaved into the large fragment HN and into a 9 K glycopeptide which is lost after cleavage. The hydrophobic sequence forming both the signal and the membrane anchor is boxed. Carbohydrate side chains are also indicated

Significance of Proteolytic Cleavage for the Pathogenicity of Rotaviruses

The infectivity of rotaviruses which do not contain an envelope is activated by trypsin cleavage of one of its capsid proteins, VP3 (Kalica et al. 1983). In genetic studies, VP3 was identified as a virulence marker (Offit et al. 1986). Since rotavirus strains differ from each other again by single arginine residues or paired basic amino acids at the cleavage site of VP 3 (López et al. 1986), it is tempting to speculate that clevability of this surface protein determines pathogenicity, although such a correlation has not been demonstrated yet. Thus, evidence is increasing that proteolytic activation of functionally important proteins may be a rather common determinant of pathogenicity.

Conclusion

Disease resulting from virus infection is a complex event, depending on the close interaction of viral and cellular factors. Through the application of biochemical and genetic methods, it is now possible to gain an insight into the molecular basis of these interactions. Thus, it has been shown that the glycoproteins of enveloped viruses play a

Table 3. Cleavage sites of the F-glycoprotein of different strains of Newcastle disease virus

Virus strain	Pathogenicity for chicken	Viral glycoproteins	Cleavage sites (F_2)	(F_1)
Italien	Velogenic	$F_{1,2}$, HN	NH_2--Gly-Gly-Arg-Arg-Gln-ARG-ARG-▼	Phe-Ile-Gly--COOH
Miyadera	Velogenic	$F_{1,2}$, HN	NH_2--Gly-Gly-Arg-Arg-Gln-ARG-ARG-▼	Phe-Ile-Gly--COOH
Herts	Velogenic	$F_{1,2}$, HN	NH_2--Gly-Gly-Arg-Arg-Gln-ARG-ARG-▼	Phe-Ile-Gly--COOH
Victoria	Velogenic	$F_{1,2}$, HN	NH_2--Gly-Gly-Arg-Arg-Gln-LYS-ARG-▼	Phe-Ile-Gly--COOH
Beaudette C	Mesogenic	$F_{1,2}$, HN	NH_2--Gly-Gly-Arg-Arg-Gln-LYS-ARG-▼	Phe-Ile-Gly--COOH
La Sota	Lentogenic	F_O, HN	NH_2--Gly-Gly-Gly-Arg-Gln-Ser-ARG-▽	Phe-Ile-Gly--COOH
D26	Lentogenic	F_O, HN_O	NH_2--Gly-Gly-Gly-Lys-Gln-Gly-ARG-▽	Leu-Ile-Gly--COOH
Queensland	Lentogenic	F_O, HN_O	NH_2--Gly-Glu-Gly-Lys-Gln-Gly-ARG-▽	Leu-Ile-Gly--COOH
Ulster	Lentogenic	F_O, HN_O	NH_2--Gly-Gly-Gly-Lys-Gln-Gly-ARG-▽	Leu-Ile-Gly--COOH

▼ Cleavage site of endoprotease recognizing pairs of basic residues.
▽ Cleavage site of endoprotease recognizing a single arginine.

Amino acids eliminated by the sequential action of endoprotease and carboxypeptidase are indicated by capitals. The sequence data of Victoria and Beaudette C strains are taken from McGinnes and Morrison (1986) and Chambers et al. (1986b), the other sequences are from Toyoda et al. (1987).

central role in the initiation of infection. They are responsible not only for the adsorption of virions to cellular receptors, but also for the entry of the genome into the cell by the fusion of viral envelopes with cellular membranes. Evidence is growing that the fusogenic glycoproteins are frequently activated by cellular protease. The structure of the proteins at the cleavage site and the availability of a suitable protease are critical for tissue tropism, spread of the virus in the infected organism and, thus, for pathogenicity.

Acknowledgments. Work by the authors was supported by the Deutsche Forschungsgemeinschaft (Sonderforschungsbereich 47) and by the Fonds der Chemischen Industrie. M.T. was a recipient of a fellowship of the Alexander von Humboldt-Stiftung.

References

Air GM, Laver WG (1986) The molecular basis of antigenic variation in influenza virus. Adv Virus Res 31:53-102

Auperin DD, Sasso DR, McCormick JB (1986) Nucleotide sequence of the glycoprotein gene and intergenic region of the Lassa virus S genome RNA. Virology 154:155-167

Binns MM, Boursnell MEG, Cavanagh D, Pappin DJC, Brown TDK (1985) Cloning and sequencing of the gene encoding the spike protein of the coronavirus JBV. J Gen Virol 66:719-726

Blumberg BM, Giorgi C, Rose K, Kolakovsky D (1985) Sequence determination of the Sendai virus fusion protein gene. J Gen Virol 66:317-331

Bosch FX, Orlich M, Klenk H-D, Rott R (1979) The structure of the hemagglutinin, a determinant for the pathogenicity of influenza viruses. Virology 95:197-207

Bosch FX, Garten W, Klenk H-D, Rott R (1981) Proteolytic cleavage of influenza virus hemagglutinins. Primary structure of the connecting peptide between HA1 and HA2 determines proteolytic cleavability and pathogenicity of avian influenza viruses. Virology 113:725-735

Chambers P, Miller NS, Emmerson PT (1986b) Nucleotide sequence of the chain gene encoding the fusion glycoprotein of Newcastle disease virus. J Gen Virol 67: 2685-2694

Collins PL, Huang AT, Wertz GW (1984) Nucleotide sequence of the gene encoding the fusion (F) glycoprotein of human respiratory syncytial virus. Proc Natl Acad Sci USA 81:7683-7687

Docherty K, Steiner DF (1982) Post-translational proteolysis in polypeptide hormone biosynthesis. Annu Rev Physiol 44:625-638

Fang R, Min-Jou W, Huylebroeck D, Devos R, Fiers W (1981) Complete structure of A/duck/Ukraine/63 influenza hemagglutinin gene: animal virus as progenitor of human H3 Hong Kong 1968 influenza hemagglutinin. Cell 25:315-323

Feldmann H (1988) The nucleotide sequence of the hemagglutinin gene of the influenza virus A/chick/Germany/49 (H10N7). Thesis, Philipps Univ, Marburg

Garoff H, Frischauf A-M, Simons K, Lehrach H, Delius H (1980) Nucleotide sequence of cDNA coding for Semliki Forest virus membrane glycoproteins. Nature (London) 288:236-241

Garten W, Berk W, Nagai Y, Rott R, Klenk H-D (1980) Mutational changes of the protease susceptibility of glycoprotein F of Newcastle Disease Virus: Effects on pathogenicity. J Gen Virol 50:135-147

Garten W, Bosch FX, Linder D, Rott R, Klenk H-D (1981) Proteolytic activation of the influenza hemagglutinin: The structure of the cleavage site and the enzymes involved in cleavage. Virology 115:361-374

Garten W, Linder D, Rott R, Klenk H-D (1982) The cleavage site of the hemagglutinin of fowl plague virus. Virology 122:186-190

Garten W, Klenk H-D (1983) Characterization of the carboxypeptidase involved in the proteolytic cleavage of the influenza haemagglutinin. J Gen Virol 64:2127-2137

Garten W, Kuroda K, Schuy W, Naruse H, Scholtissek C, Klenk H-D (1985) Haemagglutinin transport mutants. Vaccine Suppl 3:227-229

Gething MJ, White JM, Waterfield MD (1978) Purification of the fusion protein of Sendai virus: Analysis of the NH2-terminal sequence generated during precursor activation. Proc Natl Acad Sci USA 75:2737-2740

Gething MJ, Bye J, Skehel JJ, Waterfield MD (1980) Cloning and DNA sequence of double-stranded copies of hemagglutinin genes from H2 and H3 strains elucidates antigenic shift and drift in human influenza virus. Nature (London) 287:301-306

Higa HH, Rogers GN, Paulson JC (1985) Influenza virus hemagglutinins differentiate between receptor determinants bearing N-acetyl-, N-glycollyl- and N,O-diacetyl neuraminic acids. Virology 144:279-282

Homma M, Ohuchi M (1973) Trypsin action on the growth of Sendai virus in tissue culture cells. III. Structural difference of Sendai viruses grown in eggs and in tissue culture cells. J Virol 12:1457-1465

Hsu M-C, Scheid A, Choppin PW (1987) Protease activation mutants of Sendai virus: sequence analysis of the mRNA of the fusion protein (F) gene and direct identification of the cleavage-activation site. Virology 156:84-90

Huang RTC, Rott R, Klenk H-D (1981) Influenza viruses cause hemolysis and fusion of cells. Virology 110:243-247

Kalica AR, Flores I, Greenberg HB (1983) Identification of the rotaviral gene that codes for hemagglutination and protease-enhanced plaque formation. Virology 125:194-205

Kawaoka Y, Webster RG (1985) Evolution of the A/chicken/Pennsylvania/83 (H5N2) influenza virus. Virology 146:130-137

Kawaoka Y, Naeve, CW, Webster RG (1984) Is virulence of H5N2 influenza viruses in chickens associated with loss of cabohydrate from the hemagglutinin? Virology 139:303-316

Keil W, Geyer R, Dabrowski J, Dabrowski U, Niemann H, Stirm S, Klenk H-D (1985) Carbohydrates of influenza virus IV. Structural elucidation of the individual glycans of the FPV hemagglutinin by two-dimensional 1H NMR and methylation analysis. EMBO J 4:2711-2720

Klenk H-D, Rott R (1980) Cotranslational and posttranslational processing of viral glycoproteins. Curr Top Microbiol Immunol 90:19-48

Klenk H-D, Rott R, Orlich M, Blödorn J (1975) Activation of influenza A viruses by trypsin treatment. Virology 68:426-439

Klenk H-D, Rott R, Orlich M (1977) Further studies on the activation of influenza virus by proteolytic cleavage of the hemagglutinin. J Gen Virol 36:151-161

Klenk H-D, Garten W, Bosch FX, Rott R (1984) The role of the haemagglutinin as a determinant for the pathogenicity of avian influenza viruses. In: Stuart Harris C, Potter CW (eds) The molecular virology and epidemiology of influenza. Academic Press, London New York, pp 195-209

Lamb RA (1983) The influenza virus RNA segments and their encoded proteins. In: Palese P, Kingsbury DW (eds) Genetics of influenza virus. Springer Berlin Heidelberg New York, pp 21-69

Lazarowitz SG, Choppin PW (1975) Enhancement of the infectivity of influenza A and B viruses by proteolytic cleavage of the hemagglutinin polypeptide. Virology 68:440-454

Lazarowitz SG, Goldberg AR, Choppin PW (1973) Proteolytic cleavage by plasmin of the HA polypeptide of influenza virus. Host cell activation of serum plasminogen. Virology 56:172-180

López S, Arias CF, Méndez E, Espejo RT (1986) Conservation in rotaviruses of the protein region containing the two sites associated with trypsin enhancement of infectivity. Virology 154:224-227

McGinnes CW, Morrison TK (1986) Nucleotide sequence of the gene encoding the newcatle disease virus fusion protein and comparisons of paramyxovirus fusion protein synthesis. Virus Res 5:342-356

Naeve CW, Webster RG (1983) Sequence of the haemagglutinin gene from influenza virus A/Seal/Mass/1/80. Virology 129:298-308

Nagai Y, Klenk H-D (1977) Activation of precursors to both glycoproteins of Newcatle disease virus by proteolytic cleavage. Virology 77:125-134

Nagai Y, Klenk H-D, Rott R (1976) Proteolytic cleavage of the viral glycoproteins and its significance for the virulence of Newcastle disease virus. Virology 72:494-508

Offit PA, Blavat G, Greenberg HB, Clark HF (1986) Molecular basis of rota-virus virulence: role of gene segment 4. J Virol 57:46-49

Paterson RG, Harris TJR, Lamb RA (1984) Fusion protein of the paramyxovirus SV5: Nucleotide sequence of mRNA predicts a highly hydrophobic glycoprotein. Proc Natl Acad Sci USA 81:6706-6710
Porter AG, Barber C, Carey NH, Hallewell RA, Threfall G, Emtage JS (1979) Complete nucleotide sequence of an influenza virus hemagglutinin gene from cloned DNA. Nature (London) 282:471-477
Pritzer E (1985) The nucleotide sequence of the hemagglutinin gene of the fowl plague virus strain A/FPV/Dutch/27. Thesis, Justus-Liebig-Univ, Gießen
Ratner C, Haseltine W, Patarce R, Livak KJ, Starcich B, Josephs SF, Dovan ER, Rafaski JA, Whitehorn EA, Baumeister K, Ivanoff L, Petteway SR, Pearson ML, Lautenberger JA, Papas TS, Ghrayes I, Chang NT, Gallo RC, Wong-Staal F (1985) Complete nucleotide sequence of the AIDS virus, HTLV III. Nature (London) 313:277-284
Rice JM, Strauss JH (1981) Nucleotide sequence of the 26S mRNA of Sindbis virus and deduced sequence of the encoded virus structural proteins. Proc Natl Acad Sci USA 78:2062-2066
Richardson C, Hull D, Greer P, Hasel K, Berkovich A, Englund G, Bellini W, Rima B, Lazzarini R (1986) The nucleotide sequence of the mRNA encoding the fusion protein of measles virus (Edmonston strain): A comparison of fusion proteins from several different paramyxoviruses. Virology 155:508-523
Richardson CR, Scheid A, Choppin PW (1980) Specific inhibition of paramyxovirus and myxovirus replication by oligopeptides with amino acid sequences similar to those at the N-termini of F1 or HA2 viral polypeptides. Virology 105:205-222
Robertson L, Caley IP, Moore I (1958) Importance of Staphylococcus aureus in pneumonia in the 1957 epidemic of influenza A. Lancet 2:233-236
Rogers GN, Paulson JC (1983) Receptor determinants of human and animal influenza virus isolates: Differences in receptor specificity of the H3 hemagglutinin based on species of origin. Virology 127:361-373
Rogers GN, Herrler G, Paulson JC, Klenk H-D (1986) Influenza C virus uses 9-O-acetyl-N-acetylneuraminic acid as a high affinity receptor determinant for attachment to cells. J Biol Chem 261:5947-5951
Rott R, Klenk H-D (1986) Pathogenicity of influenza virus in model systems. In: Kendal AP, Patriarca PA (eds) Options for the control of influenza. Liss, New York, pp 53-62
Rott R, Reinacher M, Orlich M, Klenk H-D (1980) Cleavability of hemagglutinin determines spread of avian influenza viruses in the chorioallantoic membrane of chicken embryo. Arch Virol 65:123-133
Rott R, Orlich M, Klenk H-D, Wang ML, Skehel JJ, Wiley JC (1984) Studies on the adaptation of influenza viruses to MDCK cells. EMBO J 3:3329-3332
Scheid A, Choppin PW (1974) Identification of biological activities of paramyxovirus glycoproteins activation of cell fusion, hemolysis, and infectivity by proteolytic cleavage of an inactive precursor protein of Sendai virus. Virology 57:475-490
Scheid A, Choppin PW (1976) Protease activation mutants of Sendai: Activation of biological properties by specific proteases. Virology 69:265-277
Scheid A, Caliguiri LA, Compans RW, Choppin PW (1972) Isolation of paramyxovirus glycoproteins. Association of both hemagglutinin and neuraminidase activities with the larger SV5 glycoprotein. Virology 50:640-652
Scheid A, Graves MC, Silver SM, Choppin PW (1978) Studies on the structure and function of paramyxovirus proteins. In: Mahy BWJ, Barry RD (eds) Negative strand viruses and the host cell. Academic Press, London New York, pp 181-193
Schuy W, Garten W, Linder D, Klenk H-D (1984) The carboxyterminus of the hemagglutinin-neuraminidase of Newcastle disease virus is exposed at the surface of the viral envelope. Virus Res 1:415-426
Schwarz D, Tizard R, Gilbert W (1983) Nucleotide sequence of Rous sarcoma virus. Cell 32:853-869
Server AC, Smith JA, Waxham MN, Wolinsky JS, Goodman HM (1985) Purification and amino terminal protein sequence analysis of the mumps virus fusion protein. Virology 144:373-383
Skehel JJ, Bayley PM, Brown EB, Martin SR, Waterfield MD, White JM, Wilson JA, Wiley DC (1982) Changes in the conformation of influenza virus hemagglutinin at the pH optimum of virus-mediated membrane fusion. Proc Natl Acad Sci USA 79:968-972

Sleigh MJ, Both GW, Brownlee GG, Bender VJ, Moss BA (1980) The haemagglutinin gene of influenza A virus: nucleotide sequence analysis of cloned DNA copies. In: Laver G, Air GM (eds) Structure and variation in influenza viruses. Elsevier/ North HOlland Biomed Press, Amsterdam New York, pp 69-79

Spriggs MK, Olmsted RA, Venkatesan S, Coligan JE, Collins PL (1986) Fusion glycoprotein of human parainfluenza virus type 3: nucleotide sequence of the gene, direct identification of the cleavage-activation site, and comparison with other paramyxoviruses. Virology 152:241-251

Stuart-Harris CH, Schild GC, Oxford JS (1985) Influenza: The viruses and the disease, 2nd edn. Arnold, London

Tashiro M, Ciborowski P, Klenk H-D, Pulverer G, Rott R (1987a) Role of Staphylococcus aureus in development of influenza pneumonia. Nature (London) 325:536-537

Tashiro M, Ciborowski P, Reinacher M, Pulverer G, Klenk H-D, Rott R (1987b) Synergistic role of Staphylococcal proteases in the induction of influenza virus pathogenicity. Virology 157:421-430

Tashiro M, Klenk H-D, Rott R (1987c) Inhibitory effect of a protease inhibitor, leupeptin, on the development of influenza pneumonia, mediated by concomitant bacteria. J Gen Virol 68:2039-2041

Toyoda T, Sakaguchi T, Imai K, Inocencio NM, Gotoh B, Hamaguchi M, Nagai Y (1987) Structural comparison of the cleavage-activation site of the fusion glycoproteins between virulent and avirulent strains of Newcastle disease virus. Virology 158: 242-247

Wilson IA, Skehel JJ, Wiley DC (1981) The haemagglutinin membrane glycoproteins of influenza virus: structure at 3A resolution. Nature (London) 289:355-373

Winter G, Fields S, Brownlee GG (1981) Nucleotide sequence of the haemagglutinin of a human influenza virus H1 subtype. Nature (London) 292:72-75

Virus Genes Involved in Host Range and Pathogenicity

C. Scholtissek[1]

Introduction

The pathogenicity of a virus should always be considered in relation to its host, the organ or the individual cell in which it multiplies. During the infectious cycle, viral proteins have to cooperate with each other and have to interact with cellular components at the right time, in the correct sequence, in order to enable the virus to replicate. These cooperations and interactions can be influenced by mutations, or recombination of viral genes in the sense that pathogenic properties of a virus might be lost or created. By correlating specific changes of viral structures with such biological phenotypes, general molecular mechanisms related to pathogenicity can be studied and, when understood, may eventually be used to overcome the virus disease.

Modern gene technology allows the dissection of the genome of DNA viruses using specific restriction enzymes and the placing of the various genes into new genetic environments. The RNA of RNA viruses can be reverse transcribed into DNA, which can also be correspondingly dissected and recombined. However, viruses with a segmented genome are especially helpful in such studies, since they do not need to be dissected. They exhibit the phenomenon of reassortment in that, after infection of a cell, or an organism, with two different virus strains, the various nucleic acid segments can be mixed and replaced with each other. Since the influenza A virus system has been best studied in this respect, this article will concentrate mainly on studies with influenza A viruses and, at the end, a few corresponding data on reoviruses will be mentioned.

Influenza A Virus Genes Involved in Tropism, Pathogenicity and Species Specificity

Influenza A viruses are relatively complex RNA-containing viruses (Fig. 1), which contain two glycoproteins at the surface embedded into the lipid bilayer: the haemagglutinin (HA) and neuraminidase (NA). The membrane protein is located beneath the lipid bilayer. The nucleocapsid consists of the three polymerase proteins PB2, PB1 and PA, and the nucleoprotein (NP). The viral genome is associated with this structure. It consists of eight RNA segments (= genes) of negative polarity which have been assigned to the various gene products (Fig. 1, left) [1,2,3].

After double infection of a cell with two different influenza A virus strains, the 16 RNA segments behave like chromosomes and reassort freely (Fig. 2). Theoretically, 254 different reassortants can be

[1]Institut für Virologie, Justus-Liebig-Universität Giessen, Frankfurter Straße 107, D-6300 Giessen, FRG

38. Colloquium Mosbach 1987
Molecular Basis of Viral and Microbial Pathogenesis

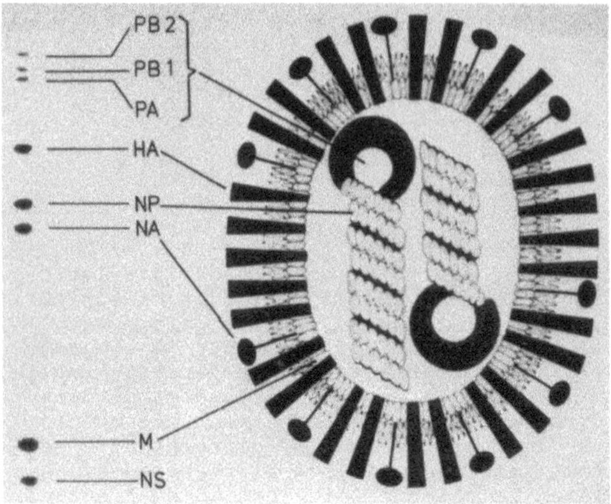

Fig. 1. Structure and structural elements of an influenza A virus (*right*) and assignment of the gene products (*middle*) to the viral RNA segments (genes) separated by polyacrylamide gel electrophoresis (*left*). The polymerase complex is located in the interior of the particle. It consists of the three P proteins *PB2*, *PB1*, and *PA* and the nucleoprotein *NP*. The lipid bilayer is covered on the inside by the membrane or *M* protein. The two glycoproteins, haemagglutinin (*HA*) and neuraminidase (*NA*), are located on the exterior surface. The smallest RNA segment contains the information for two nonstructural proteins (*NS*)

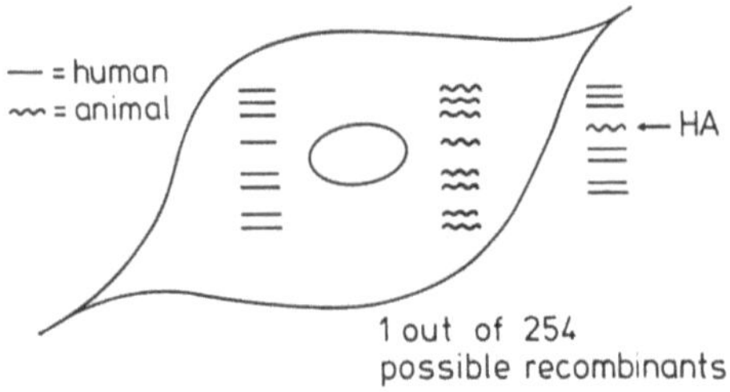

Fig. 2. Scheme to explain the creatin of a new pandemic influenza A strain (antigenic shift). A cell becomes doubly infected by an animal virus and by the prevailing human influenza A strain. One of the 254 possible new reassortants has the gene constellation shown on the right. Only the HA gene is derived from the animal strain, while all the other genes are derived from the human virus

expected. Nature takes advantage of this reassortment during the so-called antigenic shift by replacing at least the HA gene (Fig. 2). When the subtype which started the pandemic in 1968 - the H3N2 Hong Kong strain - was analyzed by the hybridization technique, it was shown that seven of the eight RNA segments were derived from the foregoing human H2N2 strain, and only the HA gene had been replaced. The "new" HA gene was genetically closely related to a HA gene of a duck H3 virus already isolated in 1963 [4]. The HA induces the production of neutralizing antibodies and thus, in 1968, such antibodies against this new H3N2 virus were not present in the human population and the Hong Kong virus with a new surface HA could start a pandemic. Since the symptoms caused by the new virus were the same, the genes responsible for the original species specificity, organ tropism etc. were taken over from the foregoing H2N2 strain.

Not all of the 254 theoretically possible reassortants are compatible for growth in a given host. Thus, some of them might multiply in chick embryo cells (CEC), but not in dog kidney (MDCK) cells. Some others might grow in MDCK cells, but not in CEC. Other reassortants again might grow in both or other cell types. This is the background for understanding how host range, organ tropism and pathogenic properties can be influenced and changed by reassortment.

In nature, "wolves in sheep's clothing" are produced during the antigenic shift. Therefore, we have tried to obtain a nonpathogenic potential live vaccine strain by constructing "a sheep in wolf's clothing".

For this purpose we used temperature-sensitive (ts) mutants of fowl plague virus (FPV), a highly pathogenic avian influenza A strain, which is released from many cells - including CEC - with a cleaved HA. By double infection of CEC with these ts mutants and other prototype influenza strains with a noncleaved HA (in CEC, see presentation by H.-D. Klenk) at the nonpermissive temperature, we selected for reassortants which carried the cleavable HA of FPV. At least the genes which carry the ts defects were replaced by the allelic genes of the prototype strains used for rescue [5]. When these specific reassortants were tested for their pathogenicity in chickens it became apparent that, for example, replacement of only RNA segment 1 (PB2 gene) of FPV by the allelic gene of the human PR8 strain was sufficient to lead to a virus completely nonpathogenic for chickens, but containing the surface antigens of the pathogenic FPV. Chickens inoculated with this synthetic reassortant withstood a superinfection with the wild-type FPV. When a swine influenza strain was used for rescue of the same ts mutant, the corresponding reassortants carrying the PB2 gene of the swine virus were as pathogenic as wild-type FPV. In order to obtain a nonpathogenic reassortant with the swine virus, we had to use, for rescue, a ts mutant of FPV with a defect in RNA segment 2 (PB1 gene) [6]. With the human Hong Kong strain as rescuing virus, replacement of the PB1 or the PB2 gene of FPV led to nonpathogenic reassortants. However, replacement of both genes concomitantly resulted in a highly pathogenic reassortant. Summarizing the results of our studies - concerning FPV in chickens - we can conclude that besides a cleavable haemagglutinin, the gene constellation - this means the correct collaboration - of the components of the polymerase complex (nucleocapsid) plays an important role for pathogenic properties of FPV. This does not exclude the possibility that the collaboration of the other gene products also contributes to pathogenicity.

If it is possible to lose, by reassortment, pathogenic properties, it should also be possible to create reassortants with a different host range or organ tropism, starting with two parent strains, which do not multiply in that host or organ. Since mice are not natural hosts for influenza viruses, most of the influenza strains freshly isolated from various hosts are nonpathogenic for mice. When reassortants between FPV and other human or avian influenza strains were tested in young mice, it could be demonstrated that certain reassortants were able to kill the animals. In contrast to the parent strains, which were nonpathogenic for mice, these new reassortants grew to high titres in the brains of infected mice as well as in brain cells cultured in vitro [7]. As shown in Table 1, which considers reassortants between FPV and another avian influenza A strain (virus N) as an example, the gene constellation of the polymerase complex (nucleocapsid) is again important [8]. This is an especially clear example of a change of organ tropism by reassortment paralleled by an increase in pathogenicity. Most importantly, immunized mice were not protected against an intranasal infection, which is the natural route of infection, by these neurotropic reassortants, since the virus was taken up by the nerve endings in the nasal mucosa and spread via the olfactory bulb over the entire brain. Immunization only prevented viremia and a generalized infection but had no influence on virus multiplication in the brain, and the animals died [9].

In nature, the creation of pathogenic influenza A viruses seems indeed to happen. In the winter 1979/80 many dead seals were found on the New England coast of the USA. From the lungs and brains of these dead animals, an influenza A virus could be isolated which was a reassortant between avian influenza viruses [10,11]. It is not yet clear whether this

Table 1. Multiplication of reassortants between fowl plague virus (F) and A/chicken/Germany "N"/49 (N) in lungs and brains after intranasal inoculation of 2-day-old mice. With exception of the first virus (virus N) the genes PB2, HA, NA and M of the reassortants are derived from fowl plague virus. If the PB1- or/and PA-gene is derived from virus N, the reassortants are neurotropic for mice. If in addition the NS-gene is derived from virus N, the virus titer found in the brain increases even further. However, replacement of the NP-gene causes loss or decrease of neurotropic properties. All animals with detectable virus in their brains die. However, the mean death time correlates inversely with the virus titer in the brains [8].

Origin of the genes				Virus titre 4 days after infection in (Plaque-forming units/ml suspension)	
PB1	PA	NP	NS	Lung	Brain
N	N	N	N	10^4	$<10^1$
F	F	F	F	10^1	$<10^1$
N	F	F	F	10^6	10^3
N	F	F	<u>N</u>	10^6	10^5
F	N	F	F	10^6	10^4
N	N	F	F	10^7	10^5
N	N	F	<u>N</u>	10^7	10^6
F	F	N	F	$<10^1$	$<10^1$
F	N	N	F	$<10^1$	$<10^1$
N	N	N	F	10^6	10^4

strain was so extremely pathogenic by its own or only in conjunction with another bacterial infection.

It is by no means an advantage for a virus to be highly pathogenic, because such a virus might be eradicated by eradication of its own host. However, influenza viruses can afford the "synthesis" of such highly pathogenic reassortants such as the seal virus (which apparently has never been isolated again), because there exists a huge reservoir of many different influenza A strains in feral water fowl which distribute the viruses all over the world. These viruses are nonpathogenic for the natural hosts, because they multiply only in a very restricted region in the gut [12]. How can we explain this fact? In the presentation by Dr. Klenk, the importance of the cleavage of the hemagglutinin for starting the infection by fusion of viral and cellular membranes in endosomes at a pH of about 5 has been discussed in detail. When influenza viruses with a cleaved HA are treated in the test tube at a pH between 5 and 6, the corresponding conformational change also occurs [13]; however, this leads to an irreversible inactivation of infectivity. If viruses with noncleaved HA are treated at low pH, this conformational change does not occur [13], and, as shown in Fig. 3, they survive and can be activated again by trypsin [14]. Influenza viruses can survive in the slightly acidic lake water [15], presumably with a noncleaved HA, and they can infect water fowls via the cloaca. The virus with the noncleaved HA can start an infection only after activation of its HA by a trypsin-like protease, which is present in the gut in exactly that region where the virus is restricted to multiply. When the virus multiplies somewhere else in the intestine, it might undergo only a single cycle of replication and

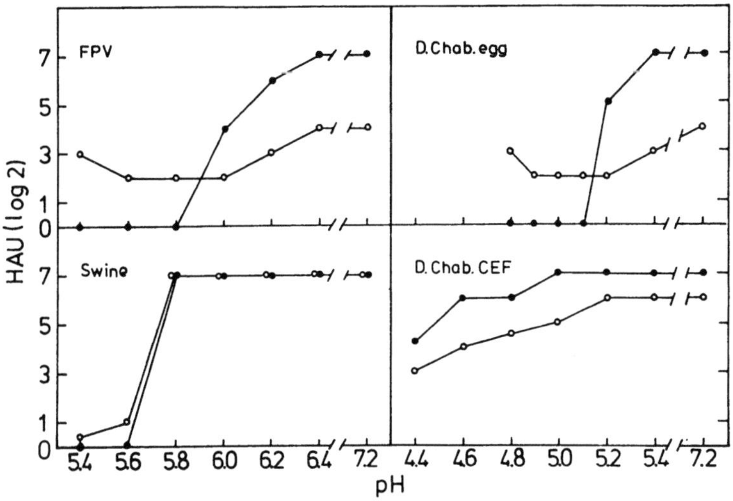

Fig. 3. pH Inactivitation of various influenza A strains. Infectious allantoic fluids of A/FPV/Rostock/34, A/swine/1976/31, or A/duck/Chabarovsk/1610/72 were adjusted to the pH as indicated on the abscissa and were incubated for 1 h at 20°C, before chick embryo cells were infected. The virus yield was determined by the HA test 7 h after infection (•). *Open circles* represent the HA titers of the infecting virus after pH treatment. Duck/Chabarovsk virus propagated on chick embryo cells having noncleaved hemagglutinin (lower right) was also investigated. For replication (o) 10 µg/ml trypsin was present in the medium

might be released with a noncleaved HA. Due to the special nitrogen excretion in birds, uric acid (pK value of 5.4) predominates in their feces. Since the virus is excreated via feces, only such particles are expected to survive in the lake water which have a noncleaved HA. Although there are certainly also other factors involved in organ tropism, by putting together these facts, namely the survival of virus with noncleaved HA in an acidic environment and activation at the site of its multiplication, the restricted organ tropism and lack of pathogenicity in water fowl and the build-up of a huge reservoir can be explained [14].

Highly pathogenic variants can be created by mutation, for example influencing the cleavage of the haemagglutinin. Such a mutation caused, for example, the outbreak of fowl plague in October 1983 in chicken farms in Pennsylvania [16]. The high mutation rate of RNA viruses allows many different viable variants to be present in a virus population, and it is only a matter of how the host selects for them. Thus, the appearance of such a highly pathogenic influenza variant in water fowl would be disastrous for the maintenance of this reservoir. However, the water fowl might again select against such highly pathogenic variants by the same mechanism as mentioned above: all variants which will multiply without the aid of external trypsin will be released from many cells with a cleaved HA - and as a rule these are the pathogenic viruses - and will be automatically destroyed during the excretion via the feces. At the moment we are testing this hypothesis by isolating nonpathogenic virus variants from a highly pathogenic population by passage at low pH.

If we look for viral genes involved in pathogenicity, we should look for viral gene products which exert pleiotropic functions like, for example, the hemagglutinin. It induces neutralizing antibodies, it reacts with the cell receptors, and it has to be cleaved in order to start the fusion between viral and cellular membranes. Such viral products which need to specifically interact with many viral and cellular compounds are especially vulnerable to mutation and to changes in genetic or cellular environments. Since we were able to artificially construct reassortants with a cleavable HA of FPV which do not multiply in chick embryo cells [17], there must exist additional factors that forbid, or allow, these viruses to multiply in a given cell. A commonly used method to modify proteins in nature and in this way to influence their functions is phosphorylation [18]. The only viral protein which is phosphorylated in all influenza A strains so far tested is the nucleoprotein (NP). As shown in Fig. 4, the phosphopeptide fingerprints of the NP's of different influenza A strains labeled in the same cell vary significantly and thus are virus-strain-specific. If, however, the NP of the same virus is labeled in different cell types, the fingerprints are also different, and this is paralleled by differences in virus yields [19]. There seems to be a subtle interplay between phosphorylation and dephosphorylation of the NP during the replication cycle (Fig. 5). In chick embryo cells infected with FPV early in the infectious cycle, when the NP accumulates in the cell nucleus [20], the phosphopeptides 2-6 of the NP can be recognized. Late in the cycle, when the NP leaves the nucleus [20] and synthesis of the M protein prevails, a new family of phosphopeptides appear and phosphopeptides 4-6 disappear. In the NP of isolated virus particles only the family of new phosphopeptides can be seen [21]. Two factors might be involved in this interplay: (1) Protein phosphokinases of different specificities might be located in different compartments of the cells, where the nucleocapsid exerts its various functions. (2) The NP might assume different conformations during the various stages in the infectious cycle, interacting with other molecules like viral RNA, the M-protein etc. We have indirect evidence that the differences in the fingerprints of, e.g., the FPV and the Hong Kong NP's might be mainly due to differences in tertiary structure of the NP's.

The question now arises as to what happens if we replace the NP gene of FPV (complex phosphopeptide fingerprint) by an NP gene of the Hong Kong strain (a simple phosphopeptide fingerprint). We have tried to do this by double infection of chicken embryo cells with ts mutants of FPV, with a defect in the NP gene, and the Hong Kong virus at the nonpermissive temperature (rescue). Since we had succeeded by using FPV mutants with defects in other genes, we were surprised to be unable to obtain plaque formers under the above-mentioned conditions after double infection with NP ts mutants and the Hong Kong virus. However, if we used dog kidney (MDCK) cells, we could isolate reassortants with the cleavable HA of FPV and the NP gene of the Hong Kong strain. The reassortants obtained in MDCK cells, however, did not form plaques in chicken embryo cells in culture, and they were nonpathogenic for chickens, although they contained the cleavable HA of FPV [17]. Here we have again an example, in which we have changed the host and pathogenic properties of a virus by reassortment.

In collaboration with K. Shortridge, who, over a period of about 15 years, has isolated many H3N2 viruses in the region of South China from different species such as humans, pigs, ducks, chickens and geese, we have tried to rescue our NP ts mutants of FPV in chicken cells. The results of Table 2 indicate that rescue of these mutants did not occur with any of the human isolates, but with all avian H3N2 viruses. The porcine isolates could be divided into two groups in this respect. Determining the genetic relatedness by the hybridization technique, it was found that the pig isolates behaving like human strains in the

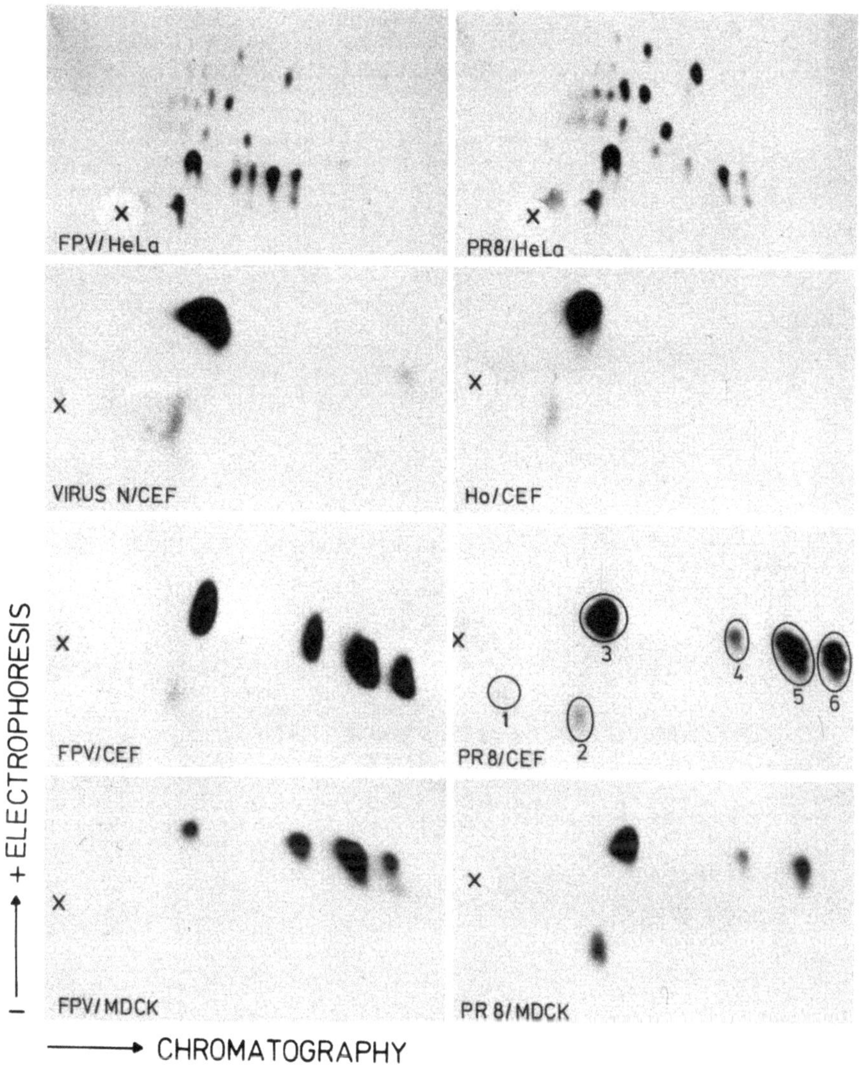

Fig. 4. Phosphopeptide fingerprints of the nucleoproteins of different influenza A strains, which were labeled with (^{32}P) orthophosphate in different cells. The following viruses were investigated: fowl plague virus (A/FPV/Rostock/34), virus *N* (A/chicken/Germany/"N"/49), *HO* (A/Hong Kong/1/68), and *PR8* (A/PR/8/34). The following host cells were investigated: *CEF* primary chick embryo fibroblasts; *MDCK* immortal canine kidney cells; *HeLa* immortal human carcinoma cells. The virus yields in MDCK cells is on average about 10% of that in CEF; the yield in HeLa cells is about 0.1% of that in CEF. During growth in HeLa cells, the nucleoprotein (NP) is overphosphorylated. During growth in MDCK cells, the PR8 NP is missing phosphopeptide 6. In FPV-infected MDCK cells, the major phosphopeptide 3 is only weakly labelled [14]

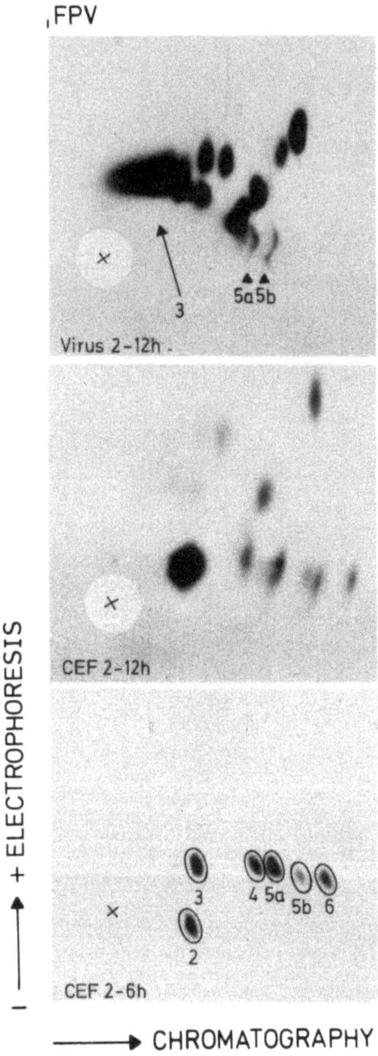

Fig. 5. Phosphopeptide fingerprints of the nucleoprotein (NP) of fowl plague virus labelled in chick embryo fibroblasts from 2 to 6 h or 2 to 12 h after infection, and of the NP isolated from purified virus particles. During late labeling, the NP isolated from cells contains new phosphopeptides, which predominate in the NP of isolated virus particles, while phosphopeptide 4, 5a, 5b, and 6 are only faintly labeled

rescue experiments had a "human-like" NP, while the other pig isolates (Sw 126 and Sw 127) had an "avian-like" NP. By doing binding studies with NP-specific monoclonal antibodies, the latter pig isolates can be differentiated from the other pig isolates as well as from the avian strains. Recent sequence studies in our laboratory are confirming these observation (Fig. 6). There exists a species-specific base sequence of the NP genes; and the NP gene of the Sw 127 is closely related to the avian NP genes, but it is not identical with them [22]. Our results are compatible with the idea that the human H3N2 viruses cannot be transmitted directly to birds, and presumably also not the other way around, without prior reassortment in pigs. Pigs seem to be more tolerant in this respect. In the example studied here, the species specificity is apparently determined by the NP gene or its gene product. The genes coding for the surface glycoproteins H3 and N2 can presumably cross the species barrier quite easily, since they were

Table 2. Summary of properties of H3N2 influenza viruses isolated from humans, pigs or birds

Property under study	Virus isolated from Human all isolates	 Pigs all except Sw126 and SW127	 Poultry all isolates	 Pigs only SW126 and Sw127
Ability to replace the NP-gene of FPV (rescue)	-	-	+	+
Hybridization against RNA segment 5 of WSN, % RNase-resistance	28-31%	28%	17%	17-18%
Binding of monoclonal antibodies against the NP of WSN; monoclones bound	3/1, 5/1, 7/3	3/1, 5/1, 7/3	3/1, 5/1, 7/3, 150/4 469/4	3/1, 5/1, 150/4

found in all isolates studied, independent of the species. The experiments on double infection with mutants with ts defects in other genes of FPV demonstrate that these genes (an exception is possible the M protein gene) also do not play an important role in determining the species specificity [23].

The particular position of the pigs as hosts for influenza A viruses of different origin might explain the sudden appearance of new strains (antigenic shift) starting pandemics in the human population. The special living conditions and the age-old agricultural techniques in the region of South Cina, allowing close contact between humans, pigs and ducks, make it understandable why most of the pandemics start from this area [24].

Concerning the influenza viral neuraminidase (NA), its involvement in certain characteristics of a virus seems to depend very much on the virus strain under investigation. The NA of the human WSN strain seems to determine its neurotropism in mice [25], while a reassortant of fowl plague virus, in which the PB2 and NA genes are derived from the human Hong Kong virus, synthesizes only traces of active neuraminidase at the body temperature of chickens (41°C). In spite of this fact, the reassortant is as pathogenic for chickens as the FPV wild type [26]. In the latter case, the NA seems not to be necessary for virus spread under the experimental conditions employed.

Reovirus Strains Involved in Tropism and Pathogenicity

Reoviruses have a segmented genome consisting of ten double-stranded RNA segments. They also exhibit the phenomenon of reassortment, at least under artificial conditions. Three of the RNA segments have a relatively large molecular weight (L), three have a medium size (M) and four a small size (S). An assignment of RNA segments to corresponding viral proteins has been achieved mainly by the group of B.N. Fields [27]. Differences in tropism and pathogenicity for mice have been found with three different reovirus types (T1, T2, T3). Only the role of three different genes, which have been especially well studied, will be mentioned here: (1) The M2 gene, whose gene product is part of the viral core, of type 1 is, after oral application, responsible for intestinal replication. The product of the M2 gene of type 3

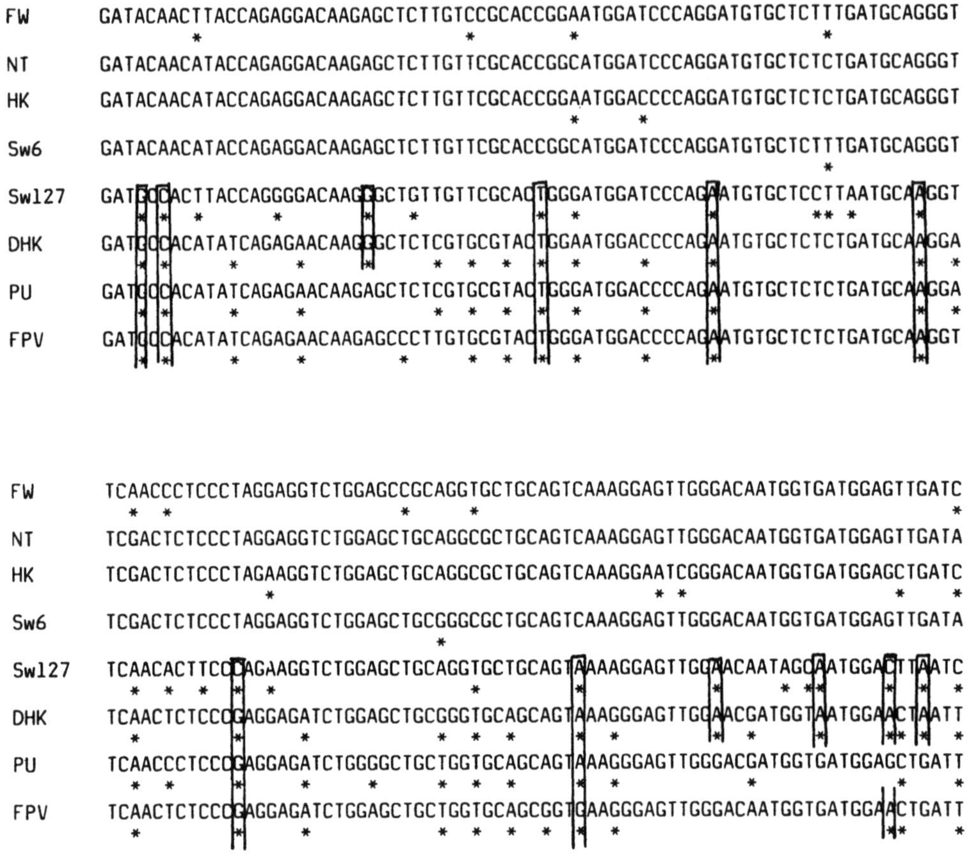

Fig. 6. Comparison of the base sequences of the nucleoprotein (NP) genes from positions 477 to 627 of the human strains A/FW/50 (H1N1), A/NT/60/68 (H3N2), and A/Hong Kong/5/83 (H3N2); of the porcine strains A/swine/Hong Kong/6/76 (H3N2), and A/swine/Hong Kong/127/82 (H3N2); and of the avian strains A/duck/Hong Kong/7/75 (H3N2), A/parrot/Ultster/73 (H7N1) and A/FPV/Rostock/34 (H7N1). *Asterisks* indicate base replacements in comparison to the human A/NT/60/68 strain

reovirus (Dearing strain) is more easily digestable by chymotrypsin. This property is regarded to be the reason for the low titres of the Dearing strain in the intestine of orally infected mice [28]. (2) The S1 gene product, which is the type-specific haemagglutinin of the outher shell, is solely responsible for the neural cell tropism of the type 3 reovirus in mice [29]. This view has been strengthened recently by the isolation of an antigenic variant of the Dearing strain using specific monoclonal antibodies, which has lost its neutrotropic properties. This loss has been achieved by a single amino acid replacement in the sigma 1 protein (haemagglutinin) [30]. (3) The sigma 3 protein (gene product of the S4 gene) is responsible for the inhibition of cellular RNA synthesis. In this respect the sigma 3 protein of the type 2 strain is more efficient than that of types 1 and 3 [31].

These few examples from studies on reoviruses demonstrate that this is also a very efficient system to study the involvement of the various gene products in tropism and pathogenicity and to elucidate the underlying mechanisms.

Summary and Concluding Remarks

Viruses with a segmented genome are especially suitable to study the involvement of specific genes or gene constellations in organ tropism, which might be directly related to pathogenicity, and in species specificity. Replacement of genes will sometimes lead only to a retardation of virus replication or of spread in an organism so that the immune system reacts fast enough to suppress any symptoms. In such cases it will be difficult to find the underlying molecular mechanisms. In the case of the influenza virus haemagglutinin, we know that its cleavability or noncleavability determines its spread, or, as with the duck viruses, its restricted multiplication in the gut. Concerning the role of the nucleoprotein, we can only guess that the subtle regulation mechanism, as revealed by multiplication cycle-dependent changes in phosphorylation and dephosphorylation, are highly vulnerable and might be severely influenced by mutation and replacement of genes whose gene products collaborate with the NP. The NP plays an important role in determining the species specificity and host range. Correspondingly, also cell components are expected to interact specifically with the NP.

Similar results have been obtained with reoviruses. Although we know that certain genes are involved in a specific tropism, the underlaying molecular mechanisms are poorly understood. Even if we can assign a certain tropism to a given gene, then, as a rule, this holds true only for this specific virus in that specific host, and a generalization is not possible.

Acknowledgments. The work performed in the author's laboratory was supported by the Deutsche Forschungsgemeinschaft, Sonderforschungsbereich 47, and by the Fonds der Chemischen Industrie. I thank Drs. R. Rott and B. Simpson for helpful discussions during the preparation of the manuscript, and Mrs. M. Seitz for typing.

References

1. Palese P (1977) The genes of influenza virus. Cell 10:1-10
2. Scholtissek C (1978) The genome of the influenza virus. Curr Top Microbiol Immunol 80:139-169
3. Scholtissek C (1986) Molecular biological background of the species and organ specificity of influenza A viruses. Angew Chem Int Ed Engl 25:47-56
4. Scholtissek C, Rohde W, Hoyningen V von, Rott R (1978) On the origin of the human influenza virus subtypes H2N2 and H3N2. Virology 87:13-20
5. Scholtissek C, Harms E, Rohde W, Orlich M, Rott R (1976) Correlation between RNA fragments of fowl plague virus and their corresponding gene functions. Virology 74:332-344
6. Scholtissek C, Rott R, Orlich M, Harms E, Rohde W (1977) Correlation of pathogenicity and gene constellation of an influenza A virus (fowl plague). I. Exchange of a single gene. Virology 81:74-80
7. Scholtissek C, Vallbracht A, Flehmig B, Rott R (1979) Correlation of pathogenicity and gene constellation of influenza A viruses. II. Highly neurovirulent recombinants derived from non-neurovirulent or weakly neurovirulent parent virus strains. Virology 95:492-500
8. Bonin J, Scholtissek C (1983) Mouse neurotropic recombinants of influenza A viruses. Arch Virol 75:255-268
9. Reinacher M, Bonin J, Narayan O, Scholtissek C (1983) Pathogenesis of neurovirulent influenza A virus infection in mice. Route of entry of virus into brain determines infection of different populations of cells. Lab Invest 49:686-692

10. Geraci R, St. Aubin DJ, Barker IK, Webster RG, Hinshaw VS, Bean WJ, Ruhnke HL, Prescott JH, Early G, Baker AS, Madoff S, Schooly RT (1982) Mass mortality of harbor seals: Pneumonia associated with influenza virus. Science 215:1129-1131
11. Hinshaw VS, Bean WJ, Webster RG (1981) Biologic and genetic characterization of an influenza A virus associated with epizootic pneumonia in seals. In: Nayak DP (ed) Genetic variation among influenza viruses. Academic Press, London New York, pp 515-524
12. Webster RG, Yakhno M, Hinshaw VS, Bean WJ, Murti G (1978) Intestinal influenza: Replication and characterization of influenza viruses in ducks. Virology 84: 268-276
13. Skehel JJ, Bayley PM, Brown EB, Martin SR, Waterfield MD, White JM, Wilson IA, Wiley DC (1982) Changes in the conformation of the influenza virus haemagglutinin at the pH optimum of virus mediated fusion. Proc Natl Acad Sci USA 79: 968-972
14. Scholtissek C (1985) Stability of infectious influenza A viruses to treatment at low pH and heating. Arch Virol 85:1-11
15. Hinshaw VS, Webster RG, Turner B (1979) Waterborn transmission of influenza A viruses? Intervirology 11:66-68
16. Kawaoka Y, Naeve CW, Webster RG (1984) Is virulence of H5N2 influenza viruses in chickens associated with loss of carbohydrate from the hemagglutinin? Virology 139:303-316
17. Scholtissek C, Koennecke I, Rott R (1978) Host range recombinants of fowl plague (influenza A) virus. Virology 91:79-85
18. Cohen P (1985) The role of protein phosphorylation in the hormonal control of enzyme activity. Eur J Biochem 151:439-448
19. Kistner O, Müller H, Becht H, Scholtissek C (1985) Phosphopeptide fingerprints of nucleoproteins of various influenza A virus strains grown in different host cells. J Gen Virol 66:465-472
20. Breitenfeld PM, Schäfer W (1957) The formation of fowl plague virus antigens in infected cells, as studied with fluorescent antibodies. Virology 4:328-345
21. Kistner O et al (in preparation)
22. Gammelin M et al (in preparation)
23. Scholtissek C, Bürger H, Kistner O, Shortridge KF (1985) The nucleoprotein as a possible major factor in determining host specificity of influenza H3N2 viruses. Virology 147:287-294
24. Shortridge KF, Stuart-Harris CH (1982) An influenza epicentre? Lancet 2:812-813
25. Sugiura A, Ueda M (1980) Neurovirulence of influenza virus in mice. I. Neurovirulence of recombinants between virulent and avirulent virus strains. Virology 101:440-449
26. Breuning A, Scholtissek C (1986) A reassortant between influenza A viruses (H7N2) synthesitzing an enzymatically inactive neuraminidase at 40° which is not incorporated into infectious particles. Virology 150:65-74
27. Ramig RF, Mustoe TA, Sharpe AH, Fields BN (1978) A genetic map of reovirus. Virology 85:531-544
28. Rubin DH, Fields BN (1980) Molecular basis of reovirus virulence: role of the M2 gene. J Exp Med 152:853-868
29. Weiner HL, Drayna D, Averill DR Jr, Fields BN (1977) Molecular basis of reovirus virulence: Role of the S1 gene. Proc Natl Acad Sci USA 74:5744-5748
30. Kaye KM, Spriggs DR, Bassel-Duby R, Fields BN, Tyler KL (1986) Genetic basis for altered pathogenesis of an immune-selected antigenic variant of reovirus type 3 (Dearing). J Virol 59:90-97
31. Sharpe AH, Fields BN (1982) Reovirus inhibition of cellular RNA and protein synthesis; role of the S4 gene. Virology 122:381-391

Molecular Heterogeneity of Pathogenic Herpes Viruses

M. Mach. H. H. Niller, and B. Fleckenstein [1]

Introduction

Herpes viruses are a large group of enveloped DNA containing animal viruses. More than 80 members of the group have been identified in various vertebrates. All herpes viruses appear identical in electron microscopy. A nucleocapsid with 162 capsomers is surrounded by an irregularly shaped membrane that is derived from the inner nuclear membrane of infected cells. It is common to all herpes viruses that, subsequent to primary infection, they persist in the host organism for life. Persisting herpes viruses can be reactivated, resulting in secondary disease. Herpes viruses are usually divided into three subgroups, based on biological criteria, such as pathogenic properties, organ tropism, and oncogenicity. Alpha-herpes viruses, including *Herpes simplex* virus (HSV) type 1 and type 2, and varizella-zoster virus (VZV) naturally persist in neural tissues. Human cytomegalovirus (HCMV) is a prototype of beta-herpes viruses. Gamma-herpes viruses are able to persist in lymphoid tissues, and some members have been shown to induce lymphoproliferative diseases. Epstein-Barr Virus (EBV) is a prototype of the gamma-1 subgroup. Two T-cell-transforming herpes viruses of non-human primates *Herpes virus (H.) saimiri* and *H. ateles*, are members of the gamma-2-subgroup.

The current taxonomy is only in part reflected by the molecular biology. Genome size variability, average GC-content, arrangement of inverted repetitions and tandem reiterations do not correlate with the biologically defined subgroups. For instance, the Marek's disease virus, a gamma-herpes virus inducing T-cell lymphoproliferations in chicken resembles some alpha- and beta-herpes viruses in sequence orientation; the genome consists of two segments that are each flanked by inverted repetitions and occur in two orientations, resulting in four isomeric forms. The T-lymphotropic gamma-herpes viruses, *H. saimiri* and *H. ateles* have a unique genome which is flanked by tandemly repeated DNA that is high in GC. In contrast, Epstein-Barr virus has several groups of internal repetitions that, in part, code for viral proteins relevant for persistence in B-cells. The rather confusing heterogeneity of herpes viruses at the molecular level led to the question whether the pathogenic phenotype of herpes viruses can be correlated in any way with certain defined properties of viral DNA or polypeptides. Honess et al. (1986) suggested plotting genetic complexity against average GC-content of protein-coding DNA-sequences in a two-dimensional diagram (Fig. 1). Alpha-herpes viruses have, in general, DNA of lower protein-coding capacity and high GC content. Viruses of the beta-subgroup and gamma-1 herpes viruses have rather large genomes. Gamma-2 herpes viruses are characterized by very low GC content. Again, various forms of isomerization patterns and repetitive sequence arrangements occur in each single subgroup. Thus, a molecular taxonomy based on overall sequence

[1] Institut für Klinische und Molekulare Virologie der Universität, Erlangen-Nürnberg, Loschgestr. 7, D-8520 Erlangen, FRG

38. Colloquium Mosbach 1987
Molecular Basis of Viral and Microbial Pathogenesis

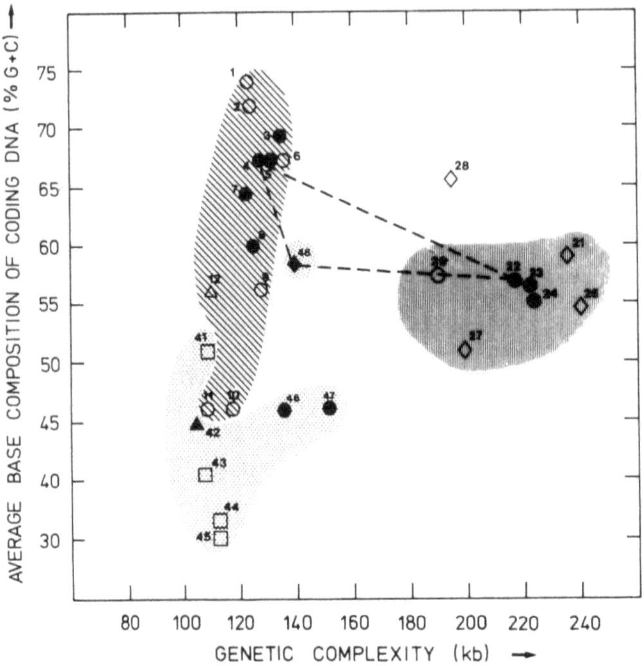

Fig. 1. Relationship between mean nucleotide composition and approximate genetic complexity for genomes of various herpes viruses

arrangements remains unsatisfactory. In this chapter we show by a few examples that the molecular genealogy of herpes viruses can be based on the comparison of local arrangement and sequence homologies of functionally equivalent genes.

Immediate Early Genes

All herpes viruses have, as far as studied, a replication cycle with a sequential order of three regulated phases of transcription and protein synthesis. Immediate early (IE) polypeptides appear to be regulatory proteins. Early proteins were often identified as virus encoded enzymes. Late proteins are essentially structural virion components. The immediate early genes and their regulatory elements give a well-documented example of herpes virus diversity. *Herpes simplex* virus has five immediate early genes (Fig. 2). The promoters and regulatory upstream sequences of four immediate early genes are located within the inverted repeats of HSV (Wagner 1985). Immediate early transcription has been shown to be dependent on activation in trans by a phosphorylated structural virion protein (Dalrymple et al. 1985). The immediate early genes of human cytomegalovirus are organized quite differently. Four immediate early transcription units are located within a continuous block of HCMV-DNA, which is part of the long unique region (Fig. 2) (Jahn et al. 1984). The major IE1-gene (Thomsen et al. 1984, Akrigg et al. 1985) and the adjacent IE2-gene (Stenberg et al. 1985) are under the control of a strong constitutive promoter/enhancer. The upstream sequences of IE-genes of *Herpes simplex* virus do not share recognizable homologies with the HCMV-enhancer. The HCMV-enhancer has been detected by its capability of replacing the 72 bp repeats of the simian virus 40 (SV40) (Boshart et al. 1985). The enhancing DNA-se-

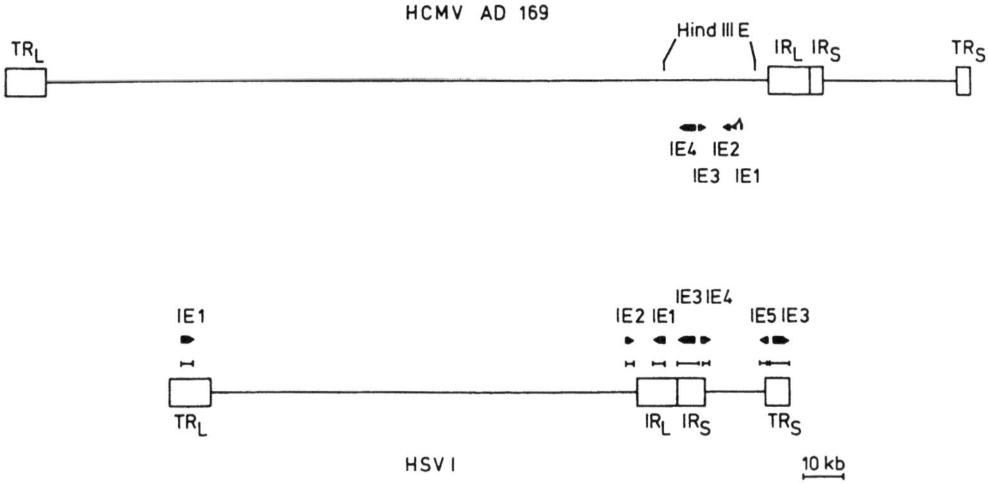

Fig. 2. Comparison of immediate early genes between HSV and HCMV

quences of HCMV extend between nucleotides -120 and -530 of the IE1 gene. The enhancer is characterized by four groups of repeats; each is represented three to five times within the upstream region of the IE1-gene. The repeats consisting of 17 bp, 18 bp, 19 bp, and 21 bp respectively are interspersed between segments of unique DNA. Each repeat has been identified by footprinting analyses as a sequence-specific protein-binding domain (Ghazal et al. 1987). Computer search for sequence homologies in the GenBank library including the VZV and EBV sequence did not identify homologies with any regulatory element of alpha- or gamma-herpes viruses (Fig. 3). Elements very similar to the palindromic 19-bp repeat unit were found only in the upstream sequences of the IE1-gene in simian and mouse cytomegalovirus. The 18-bp repeat with a protein binding domain of 11 nucleotides finds perfect matches with five repetitive elements in the mouse cytomegalovirus enhancer unit (Dorsch-Häsler et al. 1985), in the enhancer sequences of SV40 (Weiher et al. 1983, Davidson et al. 1986), and in the proviral LTR sequences from oncoviruses (Khan et al. 1982, Yoshimura et al. 1985, Ikeda et al. 1985) and human immunodeficiency virus (HIV) type 1 (Ratner et al. 1985). The 17-bp repeats of the HCMV-enhancer show nearly complete sequence homology with the hepatitis B virus enhancer only (Tognoni et al. 1985).

The structural features of beta-herpes virus enhancers exemplify the diversity of herpes viruses at the level of transcription regulation. Some protein-binding enhancer domains appear conserved between related beta-herpes viruses; however, nonrepetitive regulatory sequences and reading frames of immediate early genes are highly diverse, and no similarities were recognized with other herpes virus subgroups. The observed sequence homologies between HCMV enhancer elements and other viruses (retroviruses, hepadnaviruses) may rather be the result of evolutionary convergence of DNA-domains that are binding sites for certain cellular transcription factors.

Enzymes of Nucleotide Metabolism

The genes for some enzymes of nucleotide metabolism are well conserved between the members of different herpes virus subgroups. DNA-polymerase

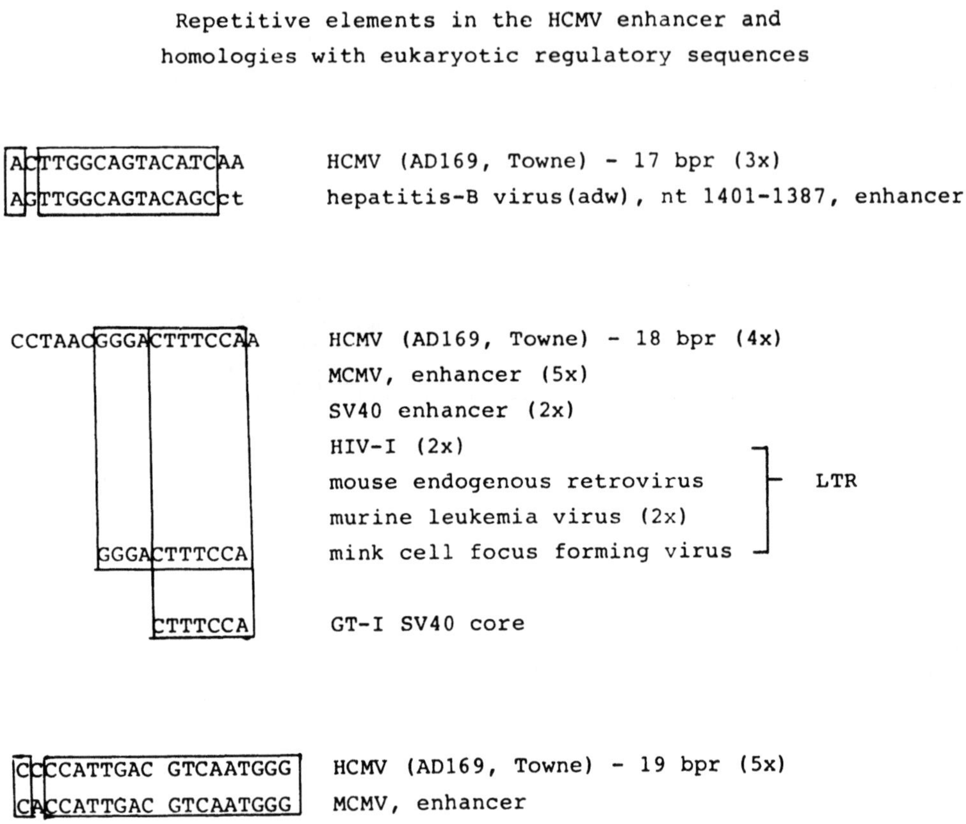

Fig. 3. Sequence homologies between the HCMV enhancer repetitive elements and other eukaryotic regulatory sequences. The protein binding domain of 11 nucleotides within the 18-bp repeat is found in various viral enhancer sequences

(Heilbronn et al. 1987, Kouzarides et al. 1987, Baer et al. 1985, Davidson and Scott 1986) and ribonucleotide reductase (Gibson et al. 1984, Dutia et al. 1986) are well documented examples. Thymidine kinase, however, is found in alpha- and in gamma-herpes viruses (Kit et al. 1974, Cheng et al. 1979, Chen et al. 1978, Honess et al. 1982); it is missing in cytomegaloviruses (Ebeling et al. 1983, Kit et al. 1975, Muller and Hudson 1977).

Thymidylate synthase was recognized as an enzyme with an unusual pattern of sequence conservation. *H. saimiri* and *H. ateles* contain a gene that is efficiently transcribed late in virus replication. Some 70 to 80% of late mRNA are represented by this transcript which is a nonspliced polyadenylated mRNA of about 2.5 kb. The mRNA has a long (about 1200 nt) 5'-nontranslated sequence with multiple start and stop codons. The single large open reading frame encodes a 33.5 kDa polypeptide (Bodemer et al. 1986). The 3'-nontranslated sequence in the *H. saimiri* mRNA is rather short; in *H. ateles*, the stop codon of the reading frame nearly coincides with the polyadenylation signal (Fig. 4). The reading frame of the overexpressed *H. saimiri* and *H. ateles* genes have significant sequence homologies with thymidylate synthase genes of human (Takeishi et al. 1985) and mouse cells (Deng et al. 1986), *E. coli* (Belfort et al. 1983), *Lactobacillus casei* (Maley et al. 1979), *Bacillus*

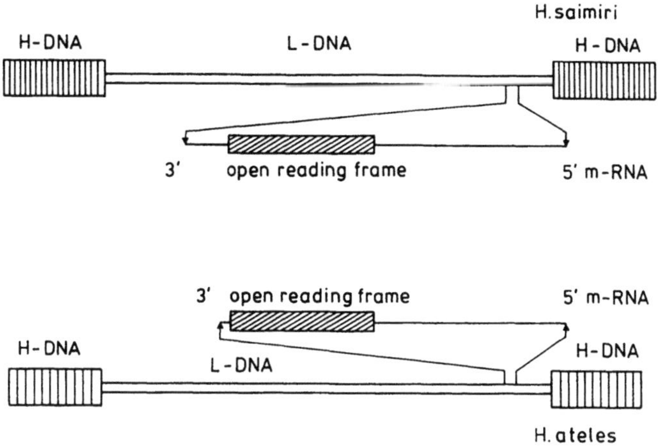

Fig. 4. Position and direction of transcription of the thymidylate synthase genes of *H. saimiri* and *H. ateles*. Both genes are located at corresponding map positions and have a long 5'-non-translated leader sequence. The 3'-nontranslated region of the *H. ateles* gene is extremely short (3 nucleotides between stop codon and polyadenylation signal)

subtilis phage 3T (Kenny et al. 1985), and bacteriophage T4 (Chu et al. 1984). Subsequent studies have indicated that the 33.5 kDa polypeptide is a functional thymidilate synthase. A homologous reading frame was found in the alpha-herpes virus VZV (Davison et al. 1986); however, not in EBV (gamma-1) (Baer et al. 1985). The enzyme was not identified in cell cultures that were lytically infected with various other herpes viruses such as *Herpes simplex* virus (alpha) (Honess et al. 1986), pseudorabies (alpha), and HCMV (beta) (Fig. 5). Apparently the presence of a thymidylate synthase gene does not correlate with taxonomic subgrouping, but rather with a high average AT-content of the genome. It may be allowed to speculate that during evolution an ancestral herpes virus may have acquired a thymidylate synthase gene from the cellular genome; herpes viruses that remained high in AT content may have preserved the gene, while it was lost during evolution to high GC-values in the majority of herpes viruses.

Structural Protein Genes

The number of surface glycoproteins in different herpes viruses and their molecular weights hint of a high phenotypic variability. However, some similarities are recognized at the genomic level (Fig. 6). As far as investigated, all human herpes viruses have reading frames for a 90-kDa glycoprotein precursor with appreciable amino acid sequence homologies (Cranage et al. 1986, Keller et al. 1986). In VZV and HCMV infected cells, this polypeptide seems to be processed by a trypsin-like cleavage. The carboxyterminal half that contains transmembrane and anchor sequence forms the immunodominant surface structure of HCMV (Mach et al. 1986). The respective polypeptide of HSV, which is not cleaved, forms the functionally important glycoprotein B. The equivalent gene product of EBV has not been characterized up to now (Pellet et al. 1985). The abundant surface polypeptide of EBV is derived from an unrelated gene.

Other structural polypeptide genes are not conserved through all subgroups. Human cytomegalovirus, for instance, has a highly immunogenic phosphoprotein of 150 kDa (Jahn et al. 1987). Phosphoproteins of *H. saimiri* and EBV that appear analogous in size and function do not reveal sequence homology on the amino acid level in comparison with HCMV; yet there is a homology between EBV and *H. saimiri* (Cameron et al. 1987). Genome positions are colinear in the two gamma-herpes viruses,

```
                     M STHTEEQHGE HQYLSQVQHI LNYGSFKNDR
                          MEELHAE HQYLSQVKHI LNCGNFKHDR
            MGDLSCWT KVPGFTLTGE LQYLKQVDDI LRYGVRKRDR
MPVAGSELPR RPLPPAAQER DAEPRPPHGE LQYLGQIQHI LRCGVRKDDR
MLVVGSEL.. ...QSDAQQL SAEAPPRHGE LQYLRQVEHI LRCGFKKEDR
                                M KQYLELMQKV LDEGTQKNDR

TGTGTLSIFG TQSRFSLENE FPLLTTKRVF WRGVVEELLW FIRGSTDSKE
TGVGTLSVFG MQSRYSLEKD FPLLTTKRVF WRGVVEELLW FIRGSTDSKE
TGIGTLSLFG MQARYNLRNE FPLLTTKRVF WRAMVEELLW FIRGSTDSKE
TGTGTLSVFG MQARYSLRDE FPLLTTKRVF WKGVLEELLW FIKGSTNAKE
TGTGTLSVFG MQARYSLRDE FPLLTTKRVF WKGVLEELLW FIKGSTNAKE
TGTGTLSIFG HQMRFNLQDG FPLVTTKRCH LRSIIHELLW FLQGDTNIAY

LSAAGVHIWD ANGSRSFLDK LGFYDRDEGD LGPVYGFQWR HFGAEYKGVG
LAASGVHIWD ANGSRSYLDK LGLFDREEGD LGPVYGFQWR HFGAEYQGLK
LAAKDIHIWD IYGSSKFLNR NGFHKRHTGD LGPIYGFQWR HFGAEYKDCQ
LSSKGVKIWH ANGSRDFLDS LGFSTREEGD LGPVYGFQWR HFGAEYRDME
LSSKGVRIWD ANGSRDFLDS LGFSARQEGD LGPVYGFQWR HFGAEYKDMD
LHENNVTIWD EWADEN.... ........GD LGPVYGKQWR AWPTP.....

RDYKGEGVDQ LKQLIDTIKT NPTDRRMLMC AWNVSDIPKM VLPPCHVLSQ
HNYGGEGVDQ LKQIINTIHT NPTDRRMLMC AWNVLDVPKM ALPPCHVLSQ
SNYLQQGIDQ LQTVIDTIKT NPESRRMIIS SWNPKDIPLM VLPPCHTLCQ
SDYSGQGVDQ LQRVIDTIKT NPDDRRIIMC AWNPRDLPLM ALPPCHALCQ
SDYSGQGVDQ LQKVIDTIKT NPDDRRIIMC AWNPKDLPLM ALPPCHALCQ
...DGRHIDQ ITTVLNQLKN DPDSRRIIVS AWNVGELDKM ALAPCHAFFQ

FYVCDGKLSC QLYQRSADMG LGVPFNIASY SLLTCMIAHV TNLVLGEFIH
FYVCDGKLSC QLYQRSADMG LGVPFNIASY SLLTCMIAHV TDLVPGEFIH
FYVANGELSC QVYQRSGDMG LGVPFNIAGY ALLTYIVAHV TGLKTGDLIH
FYVVNSELSC QLYQRSGDMG LGVPFNIASY ALLTYMIAHI TGLKPGDFIH
FYVVNGELSC MLYQRSCDMG LGVPFNIASY ALLTILIAKA TGLRPGELVH
FYVADGKLSC QLYQRSCDVF LGLPFNIASY ALLVHMMAQQ CDLEVGDFVW

TIGDAHIYVD HIDALKMQLT RTPRPFPTLR FARNVSCIDD FKADDIILEN
TLGDAHVYVN HVDALTEQLT RTPRPFPTLK FARKVASIDD FKANDIILEN
TMGDAHIYLN HIDALKVQLA RSPKPFPCLK IIRNMTDIND FKWDDFQLDG
TLGDAHIYLN HIEPLKIQLQ REPRPFPKLR ILRKVEKIDD FKAEDFQIEG
TLGDAHVYRN HIEPLKIQLQ REPRPFPKLK ILRKVETIDD FKVEDFQIEG
TGGDTHLYSN HMDQTHLQLS REPRPLPKLI IKRKPESIFD YRFEDFEIEG

YNPHPIIKMH MAV      Herpesvirus saimiri
YNPYPSIKMP MAV      Herpesvirus ateles
YNPHPPLKME MAL      Varicella Zoster Virus
YNPHPTIKME MAV      Human
YNPHPTIKME MAV      Mouse
YDPHPGIKAP VAI      Escherichia coli
```

Fig. 5. Comparison of thymidylate synthase proteins of different species. Alignments of the amino acid sequence of the *H. saimiri* thymidylate synthase with sequences of *H. ateles*, varicella-zoster-virus, human, mouse and *E. coli* thymidylate synthases. Residues that are identical in four or more sequences in the alignments shown are boxed

and the direction of transcription is identical. This example can be taken as evidence that gamma-1- and gamma-2-herpes viruses are closer related to each other than to the alpha- or beta-subgroups.

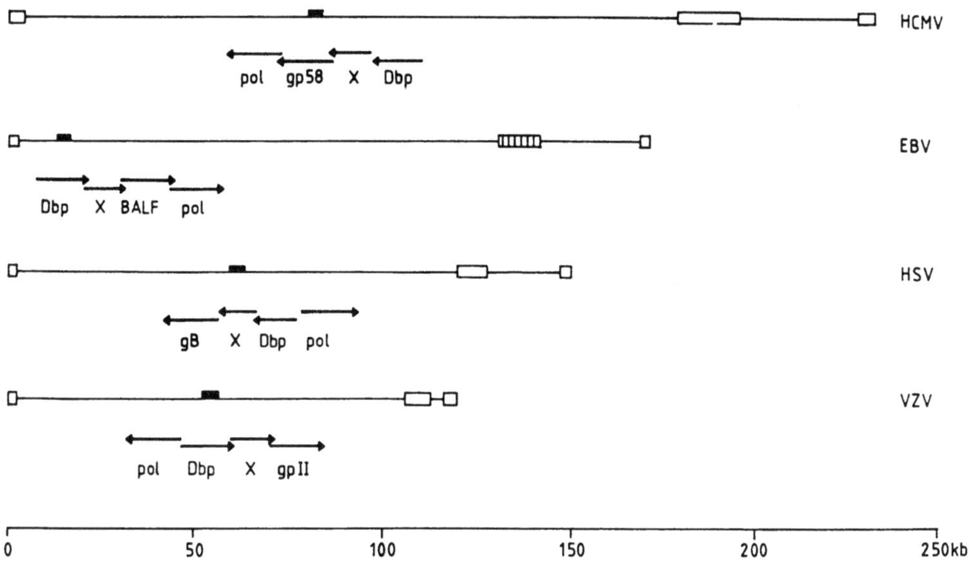

Fig. 6. Genomic arrangement of the genes coding for the DNA-polymerase and the glycoprotein B homologue in the human herpes viruses. *pol* DNA-polymerase; *Dbp* DNA-binding protein; *x* unknown reading frame; *gp58*, *BALF*, *gB*, *gpII* glycoprotein B homologue

Concluding Remarks

The pathogenic herpes viruses are highly diverse with respect to conservation or divergence of functionally equivalent genes, sequence arrangements, and deletions. Examples of conserved genes are the coding sequences for DNA-polymerases and a major surface glycoprotein. Thymidylate synthase is highly conserved in a few herpes viruses with high AT-DNA, however missing in others. A phosphorylated matrix protein allows the recognition of a relationship between the gamma-1- and gamma-2-subgroups, but shows no conservation beyond the subgroup. Regulatory proteins and promoter/enhancer-sequences are highly heterogeneous, even in close relatives. It suggests that the evolutionary adaptation and changing patterns of cell tropism have evolved primarily through modification of transcription regulating sequences, regulatory proteins and through posttranslational modification in structural proteins.

References

Akrigg A, Wilkinson GWG, Oram JD (1985) The structure of the major immediate early gene of human cytomegalovirus strain AD169. Virus Res 2:107-121

Baer R, Bankier AT, Biggin MD, Deininger PL, Farrell PJ, Gibson TJ, Gatful G, Hudson GS, Satchwell C, Tuffnell PS, Barrell BG (1985) DNA-sequence and expression of the B95-8 Epstein-Barr virus genome. Nature (London) 310:207-211

Belfort M, Maley G, Pedersen-Lane J, Maley F (1983) Primary structure of the *Escherichia coli thy* A gene and its thymidylate synthase product. Proc Natl Acad Sci USA 80:4914-4918

Bodemer W, Niller HH, Nitsche N, Scholz B, Fleckenstein B (1986) Organization of the thymidylate synthase gene of herpesvirus saimiri. J Virol 60:114-123

Boshart M, Weber F, Jahn G, Dorsch-Häsler K, Fleckenstein B, Schaffner W (1985) A very strong enhancer is located upstream of an immediate early gene of human cytomegalovirus. Cell 41:521-530

Cameron K, Stamminger Th, Craxton M, Bodemer W, Fleckenstein B, Honess RW (1987) The 160K virion protein encoded at the "right" end of the *Herpesvirus saimiri* genome is homologous to the 140K "membrane" antigen encoded at the "left" end of the Epstein-Barr virus genome. J Virol 61:2063-2070
Chen ST, Estes JE, Huang E-S, Pagano JS (1978) Epstein-Barr virus-associated thymidine kinase. J Virol 26:203-208
Cheng YC, Tsou TY, Hackstadt T, Mallavia LP (1979) Induction of thymidine kinase and DNase in varicella-zoster virus-infected cells and kinetic properties of the virus-induced thymidine-kinase. J Virol 31:172-177
Chu FK, Maley GF, Maley F, Belfort M (1984) Intervening sequence in the thymidylate synthase gene of bacteriophage T4. Proc Natl Acad Sci USA 81:3049-3053
Cranage MP, Kouzarides T, Bankier AT, Satchwell S, Weston K, Tomlinson P, Barrell B, Hart M, Bell SE, Minson AC, Smith GL (1986) Identification of the human cytomegalovirus glycoprotein B gene and induction of neutralizing antibodies via its expression in recombinant vaccinia virus. EMBO J 5:3057-3063
Dalrymple MA, McGeoch DJ, Davison AJ, Preston CM (1985) DNA sequence of the herpes simplex virus type 1 gene whose product is responsible for transcriptional activation of immediate early promoters. Nucleic Acids Res 13:7865-7879
Davidson I, Fromental C, Augereau P, Wildeman A, Zenke M, Chambon P (1986) Cell type specific protein binding to the enhancer of simian virus 40 in nuclear extracts. Nature (London) 323:544-548
Davison AJ, Scott JE (1986) The complete DNA sequence of varicella-zoster virus. J Gen Virol 67:1759-1816
Deng TL, Li DW, Jenk CH, Johnson LF (1986) Structure of the gene for mouse thymidylate synthase. Locations of introns and multiple transcription sites. J Biol Chem 261:16000-16005
Dorsch-Häsler K, Keil GM, Weber F, Jasin M, Schaffner W, Koszinowski UH (1985) A long and complex enhancer activates transcription of the gene coding for the highly abundant immediate early mRNA in murine cytomegalovirus. Proc Natl Acad Sci USA 82:8325-8329
Dutia BM, Frame MC, Subak-Sharpe JH, Clark WN, Marsden HS (1986) Specific inhibition of herpes virus ribonucleotide reductase by synthetic peptides. Nature (London) 321:439-441
Ebeling A, Keil G, Nowak B, Fleckenstein B, Berthelot N, Sheldrick P (1983) Genome structure and virion polypeptides of the primate herpes viruses *herpesvirus aotus* types 1 and 3: comparison with human cytomegalovirus. J Virol 45:115-126
Ghazal P, Lubon H, Fleckenstein B, Hennighausen L (1987) Binding of transcription factors and the creation of a large nucleoprotein complex on the human cytomegalovirus enhancer. Proc Natl Acad Sci USA (in press)
Gibson T, Stockwell P, Ginsburg M, Barrell B (1984) Homology between two EBV early genes and HSV ribonucleotide reductase and 38 K genes. Nucleic Acids Res 12: 5087-5099
Heilbronn R, Jahn G, Bürkle A, Freese U-K, Fleckenstein B, zur Hausen H (1987) Genomic localization, sequence analysis, and transcription of the putative human cytomegalovirus DNA polymerase gene. J Virol 61:119-124
Honess RW (1984) Herpes simplex and "the herpex complex": diverse observations and a unifying hypothesis. J Gen Virol 65:2077-2107
Honess RW, O'Hare P, Young D (1982) Comparison of thymidine kinase activities induced in cells productively infected with *Herpesvirus saimiri* and herpes simplex virus. J Gen Virol 58:237-249
Honess RW, Bodemer W, Cameron KR, Niller HH, Fleckenstein B, Randall RE (1986) The A+T-rich genome of *Herpesvirus saimiri* contains a highly conserved gene for thymidylate synthase. Proc Natl Acad Sci USA 83:3604-3608
Ikeda H, Laigret F, Martin MA, Repaske R (1985) Characterization of a molecularly cloned retroviral sequence associated with Fv-4 resistance. J Virol 55:768-777
Jahn G, Knust E, Schmolla H, Sarre Th, Nelson JA, McDougall JK, Fleckenstein B (1984) Predominant immediate-early transcripts of human cytomegalovirus AD169. J Virol 49:363-370
Jahn G, Kouzarides T, Mach M, Scholl B-C, Plachter B, Traupe B, Preddie E, Satchwell SC, Fleckenstein B, Barrell BG (1987) Map position and nucleotide sequence of the gene for the large structural phosphoprotein of human cytomegalovirus. J Virol 61:1358-1367

Keller PM, Davison AJ, Lowe R, Bennett C, Ellis R (1986) Identification and structure of the gene encoding gpII, a major glycoprotein of varicella-zoster virus. Virology 152:181-191
Kenny E. Atkinson T, Hartley BS (1985) Nucleotide sequence of the thymidylate synthase gene (thyP3) from the *Bacillus subtilis* phage 3T. Gene 34:335-342
Khan AS, Repaske R, Garon CF, Chan HW, Rowe WP, Martin MA (1982) Characterization of proviruses cloned from mink cell focus-forming virus-infected cellular DNA. J Virol 41:435-448
Kit S, Leung W-C, Jorgensen G, Trkula D, Dubbs DR (1974) Thymidine kinase isozymes of normal and infected cells. Cold Spring Harbor Symp Quant Biol 39:703-715
Kit S, Leung W-C, Jorgensen G, Trkula D, Dubbs DR (1975) Viral-induced thymidine kinase. Prog Med Virol 21:13-34
Kouzarides T, Bankier AT, Satchwell SC, Weston K, Tomlinson P, Barrell BG (1987) Sequence and transcription analysis of the human cytomegalovirus DNA polymerase gene. J Virol 61:125-133
Mach M, Utz U, Fleckenstein B (1986) Mapping of the major glycoprotein gene of human cytomegalovirus. J Gen Virol 67:1461-1467
Maley GF, Bellisario RL, Guarino DU, Maley F (1979) The primary structure of *Lactobacillus casei* thymidylate synthetase. J Biol Chem 254:1301-1304
Muller MT, Hudson JB (1977) Thymidine kinase activity in mouse 3T3 cells infected with murine cytomegalovirus (MCV). Virology 80:430-433
Pellet PE, Biggin MD, Barrell B, Roizman B (1985) Epstein-Barr virus genome may encode a protein showing significant amino acid and predicted secondary structure homology with glycoprotein B of herpes simplex virus 1. J Virol 56:807-813
Ratner L, Haseltine W, Patarca R, Livak KJ, Starcich B, Josephs SF, Doran ER, Rafalski JA, Whitehorn EA, Baumeister K, Ivanoff L, Petteway SR, Pearson ML, Lautenberger JA, Papas TS, Ghrayeb J, Chang NT, Gallo RC, Wong-Staal F (1985) Complete nucleotide sequence of the AIDS virus, HTLV III. Nature (London) 313: 277-284
Stenberg RM, Witte PR, Stinski MF (1985) Multiple spliced and unspliced transcripts from human cytomegalovirus immediate-early region 2 and evidence for a common initiation site within immediate-early region 1. J Virol 56:665-675
Takeishi K, Kaneda S, Ayusama D, Shimizu K, Gotoh O, Seno T (1985) Nucleotide sequence of a functional cDNA for human thymidylate synthase. Nucleic Acids Res 13:2035-2043
Thomsen DR, Stenberg RM, Goins WF, Stinski MF (1984) Promoter-regulatory region of the major immediate early gene of human cytomegalovirus. Proc Natl Acad Sci USA 81:659-663
Tognoni A, Cattaneo R, Serfling E, Schaffner W (1985) A novel expression selection approach allows precise mapping of the hepatitis B virus enhancer. Nucleic Acids Res 13:7457-7472
Wagner EK (1985) Individual HSV transcripts: characterization of specific genes. In: Roizman B (ed) The herpesviruses, vol 3. Plenum, New York
Weiher H, König M, Gruß P (1983) Multiple point mutations affecting the simian virus 40 enhancer. Science 219:626-631
Yoshimura FK Davison B, Chaffin K (1985) Murine leukemia virus long terminal repeat sequences can enhance gene activity in a cell-type-specific manner. Mol Cell Biol 5:2832-2835

Recombination of Foreign (Viral) DNA with the Host Genome Studies in Vivo and in a Cell-Free System

W. Doerfler, A. Spies, R. Jessberger, U. Lichtenberg, C. Zock, and T. Rosahl[1]

Introduction

In viral pathogenesis, the persistence of the viral genome or of certain of its segments has long been recognized as one of the predominant features of latent viral infections. This aspect of virus-host interactions has received attention in particular for the group of viruses that are thought to play a role in experimental or natural oncogenesis. Of course, other viral systems have also been investigated in this respect. In persistent or latent virus infections, the intracellular fate of the viral genome - its physical state and the factors controlling its replication and expression - has posed one of the paramount questions in studies on the interplay between virus and host cell. Our own laboratory has been fascinated with this problem for over two decades (Doerfler 1968, 1970). Studies on the integrated state of viral DNA have been considered of interest not only for viral oncogenesis but, rather in a much wider context, as a model for the insertion of foreign DNA into the genome of mammalian cells and for the mechanisms of recombination in mammalian cell systems. Perhaps the fixation of foreign DNA in cells has also played an important role in evolution.

The fixation of viral genes in the afflicted host cell is one of the preconditions for viral persistence. In the case of DNA viruses or of retroviruses, the viral DNA can be integrated into the host genome. In some instances, the viral genome can also persist in an episomal state. This latter way of persistence requires a highly coordinated replication system that has to function in synchrony with that of the host cell. From a geneticist's point of view, the insertion of foreign (viral) DNA into the host genome acquires relevance mainly for the consequences on the function of host genes. Depending on the site(s) of viral DNA integration, problems of insertional mutagenesis and effects of integrated viral genes and viral genetic signals (e.g., enhancer sequences) on adjacent host genes deserve detailed investigations. At a time when studies on the molecular mechanism of recombination have attained an important place among research topics in molecular biology, work on insertional recombination with viral genomes as model appears to offer certain advantages.

In this report, a survey of previous findings on the integration of adenovirus DNA in hamster, mouse, and human cells will precede summaries of current work: (1) The transcriptional activity of cellular DNA appears to render transcribed segments of the cellular genome more prone for insertional recombination. (2) Of special interest are those cellular sites of adenovirus DNA integration from which the foreign DNA can be reexcised. (3) More recently, we have developed systems to investigate the mechanism(s) of recombination between adenovirus and hamster DNA in a cell-free system. A cloned preinsertion site from

[1] Institut für Genetik, Universität Köln, D-5000 Köln, FRG

38. Colloquium Mosbach 1987
Molecular Basis of Viral and Microbial Pathogenesis

hamster DNA, that corresponds to an actual site of adenovirus type 12 (Ad12) DNA integration in tumor cell line CLAC1 (Stabel and Doerfler 1982), has been employed in these studies. In the present communication, we have summarized some of the current research on the topics of recombination and integration conducted in our laboratory.

A Survey of Characteristics of Integration Sites of Adenovirus DNA in the Mammalian Genome

In an attempt to reconstruct some of the essential features of the mechanism of adenovirus DNA integration in mammalian cells, we have studied in considerable detail the structure of eight sites of junction between adenovirus DNA and host cell DNA. The results of these studies have been published earlier and have been summarized in several reviews (Doerfler 1982, Doerfler et al. 1983, 1985). The synopsis in Table 1 lists the most important findings from this series of investigations. For the ease of presentation of a large number of findings, the problem has been subdivided into two divisions: Aspects of integrated adenovirus DNA and of the cellular site of integrative recombination (Table 1).

Viral DNA. The patterns of viral DNA integration appear to depend on the permissivity of the host-adenovirus systems. In hamster cells, Ad12 DNA is often integrated (nearly) intact. Hamster cells are absolutely nonpermissive for Ad12 in that viral DNA cannot replicate and late genes are not transcribed, hence persistence of the intact viral DNA cannot lead to its replication and to the killing of the host cell. In contrast, adenovirus type 2 (Ad2) replicates in hamster cells. Therefore, persistence and integration of the intact Ad2 molecule seems to be selected against, since these cells would be eliminated due to viral replication and cell killing. Ad2 DNA has always been found to be integrated in fragments or in molecules carrying minor or major deletions. This interpretation of basic differences in Ad2 and Ad12 DNA integration patterns is consistent with the biology of these systems. Obviously, there could be additional, more complex reasons for differences in integration patterns. It should also be mentioned that in some hamster tumors, Ad12 DNA fragments have been found to be integrated (Kuhlmann et al. 1982).

The site of recombination between viral and cellular DNA has frequently been found at or close to the termini of Ad2 or Ad12 DNA. Perhaps DNA termini have an inherent propensity to undergo recombination with other DNA molecules. With the adenovirus chromosome having the terminal protein covalently attached at either 5'terminus (Robinson et al. 1973, Rekosh et al. 1977), it has frequently been speculated that this terminal protein may have a role in adenovirus DNA integration. Available data justify keeping this idea alive; hopefully, it can be critically tested in the future.

Most adenovirus-transformed cell lines, adenovirus type 12-induced tumors, or cell lines established from these tumors carry multiple copies of the viral genome in an integrated form. Free copies of viral DNA persisting in an episomal state have not been observed. Persistence of single or few copies of viral DNA appears rather to be the exception. A true tandem arrangement of viral DNA molecules, in the sense of adenovirus molecule being followed by an identical molecule, has not been found. Individual molecules are frequently separated by nonviral or rearranged viral DNA sequences (Stabel et al. 1980, Kuhlmann and Doerfler 1982).

Table 1. Insertion of adenovirus DNA into the host genome[a]

1. *Viral DNA*

i. Dependence on permissivity
Ad12 DNA - frequently intact
Ad2 DNA - internal deletions; fragments integrated

ii. Site of recombination at or close to termini of viral DNA (terminal protein)

iii. Multiple copies (1 to > 30) of viral DNA per cell genome

iv. Rarely true tandems of adenovirus DNA
Interspersed cellular or rearranged viral sequences
Amplifications of terminal sequences

v. Frequently, not invariably, deletions of terminal nucleotides (8-174 nucleotides)

vi. Viral DNA associated with specific chromosomes, however, more than one site of chromosomal location

vii. Loss of viral genomes observed giving rise to revertants
Total loss of 1/2 to 1 out of 22 copies of Ad12 DNA preserved (cell line T637)

viii. Loss of viral DNA compatible with oncogenic phenotype

ix. In addition to intact Ad12 genome(s), dispersed fragments of Ad12 DNA are integrated

2. *Cellular DNA*

i. No unique specific cellular integration sequence apparent

ii. Patch homologies observed - not always found

iii. Integration into unique or repetitive cellular DNA sequences

iv. Deletion of cellular DNA (1.5 to 1.6 kb) observed or integration without loss of single cellular nucleotide

v. Specific genes in vicinity of integration site in some cases - 4.5S RNA-Endogenous intracisternal A particle (IAP) genome

vi. Transcriptional activity of cellular DNA at or close (350 nt.) to site of adenovirus DNA integration

[a]This synopsis presents a summary of the main molecular features of at least eight different sites of junction between adenovirus DNA and cellular DNA. The detailed description of our findings has been published elsewhere (Deuring et al. 1981a, Stabel and Doerfler 1982, Gahlmann et al. 1982, Gahlmann and Doerfler 1983, Deuring and Doerfler 1983, Doerfler et al. 1983, Schulz and Doerfler 1984, Lichtenberg et al. 1987, Schulz et al. 1987, R. Jessberger, S. Stabel, and W. Doerfler, manuscript in preparation).

In addition to viral DNA molecules which are integrated nearly intact or with certain internal deletions, dispersed fragments of viral DNA can also be detected. In one instance, such a dispersed fragment of Ad12 DNA has been precisely mapped to the vicinity of an integrated intact Ad12 DNA molecule (Lichtenberg et al. 1987; see below). The terminal nucleotides of the integrated Ad2 or Ad12 DNA molecules are often deleted. These deletions range in extent from 8 to 174 base pairs (Doerfler et al. 1983). There is one example in cell line HA12/7, in which not a single viral nucleotide has been missing at the transition site of viral to cellular DNA (Schulz et al. 1987).

It has been demonstrated that the integrated Ad12 DNA molecules are associated with certain hamster chromosomes. In cell line T637, which contains about 20 to 22 copies of viral DNA, more than one hamster chromosome can be shown by in situ hybridization experiments to carry Ad12 DNA molecules (Vogel et al. 1986). In particular, from the Ad12-transformed cell line T637, morphological revertants have been isolated. These revertants are characterized by the complete or partial loss of the 20 to 22 integrated Ad12 genomes. In some of the revertant cell lines, only 1/2 or 1 Ad12 DNA molecule persists in the cellular genome (Groneberg et al. 1978, Eick and Doerfler 1982). Loss of integrated viral DNA appears an infrequent but clearly demonstrable event, which seems to occur more frequently at relatively early times after the insertion of viral DNA into the host chromosome than in cell lines established in culture for a long time. It is remarkable that in the Ad12-transformed hamster cell line T637 with about 20 to 22 copies of chromosomally integrated viral DNA most of these copies can be lost simultaneously (Groneberg et al. 1978), although these viral DNA molecules are located on different chromosomes (Vogel et al. 1986). Perhaps the excision mechanism recognizes very specific signals at the insertion sites. For at least one Ad12-induced hamster tumor line it has been shown that the apparent total loss of Ad12 DNA from this cell line is compatible with the persistence of the oncogenic phenotype (Kuhlmann et al. 1982). This finding is consistent with a postulated hit-and-run mechanism of viral oncogenesis, but could also be explained by the, as yet unproven, persistence of viral genetic signals, e.g., enhancers, whose persistence cannot easily be demonstrated, but which may continue to influence regulatory mechanisms of the host. Unquestionably, the excision mechanism offers interesting possibilities for further research. This mechanism may also be of practical importance since excision could be considered a defense mechanism of the cell against the intrusion of foreign genetic material into the host genome. If this mechanism were more completely understood, perhaps means could be developed to counter the establishment of latent viral infections. The existence of an excision mechanism as a host defense has not previously attained sufficient attention.

Cellular DNA. Cellular nucleotide sequences at eight different sites of adenovirus DNA integration have been determined (Doerfler et al. 1983, Lichtenberg et al. 1987, Schulz et al. 1987) and compared with each other (Schellner et al. 1986). A specific cellular nucleotide sequence at which adenovirus DNA integrates has not been detected. Patchy homologies between viral and cellular DNA's at the sites of recombination exist and are sometimes very striking (Stabel and Doerfler 1982), but are not always found. Such patch homologies may assume an important function in the stabilization of recombination complexes as intermediates in integrative recombination.

Adenovirus DNA can integrate into unique or repetitive cellular DNA sequences. At the site of viral DNA insertion, deletions of cellular DNA sequences have been seen in one instance in the Ad12-induced mouse tumor CBA-12-1-T (Schulz and Doerfler 1984). At a different site, in the Ad2-transformed hamster cell line HE5, not a single cellular nucleotide has been obliterated (Gahlmann and Doerfler 1983). In two examples, known cellular or endogenous retroviral genes have been detected in the cellular vicinity of integrated Ad12 DNA molecules. In the tumor cell line CLAC1, the gene for a 4.5S RNA has been found in close proximity to integrated Ad12 DNA (Schulz et al. 1987). In the Ad12-induced hamster tumor T1111(2), the long terminal repeat sequence of an intracisternal A particle (IAP) sequence has been detected (Lichtenberg et al. 1987).

Foreign DNA Insertion and Transcriptional Activity at Sites of Integration

In the previous section, phenomenological details but no apparent common structural or functional parameters were described for at least eight independently cloned and analyzed integration sites of adenovirus DNA in hamster, mouse, or human DNA. For many of these insertion sites, the corresponding preinsertion sites from cells devoid of viral DNA molecules have also been cloned and their nucleotide sequences have been determined. The cloned preinsertion sites from hamster, mouse, or human cells have been used to determine transcriptional activities at these sites in the non-virus-transformed cells (Gahlmann et al. 1984, Schulz et al. 1987). The preinsertion sequences employed in these studies corresponded to sites of junction from the Ad2-transformed hamster cell line HE5 (Gahlmann et al. 1982, 1984, Gahlmann and Doerfler 1983), from the Ad12-induced mouse tumor CBA-12-1-T (Starzinski-Powitz et al. 1982, Schulz and Doerfler 1984), from the Ad12-induced hamster tumor cell line CLAC1 (Stabel and Doerfler 1982), and from the symmetric recombinant SYREC2 (Deuring et al. 1981b, Deuring and Doerfler 1983). In some experiments, a cloned integration site from the Ad12-transformed hamster cell line HA12/7 (Schulz et al. 1987, R. Jessberger, S. Stabel, and W. Doerfler manuscript in preparation) was used. The results of these analyses demonstrate that the cellular DNA sequences involved in linkage to viral DNA at these five different sites from three different species are transcribed into RNA's even in cells which have not been transformed or infected by adenoviruses. Some of these RNA's are cytoplasmic and not polyadenylated. Human cellular DNA sequences at the junction to Ad12 DNA in SYREC2 DNA are transcribed into polyadenylated cytoplasmic RNA which can be translated in vitro into a 22.4 kDa polypeptide. Hence, this human cellular sequence possibly constitutes a functional gene. The preinsertion site from a sixth junction site derived from the tumor T1111(2) has also been shown to be transcriptionally active, although the transcripts derive from sequences a few hundred nucleotides remote from the site of Ad12 DNA insertion (Lichtenberg et al. 1987; see below). The schemes in Fig. 1 present a summary of the transcriptionally active preinsertion sites and of their precise locations relative to the site of integrative recombination with adenovirus DNA. These results are consistent with the notion that the cellular DNA sequences at sites of recombination with adenovirus (foreign) DNA are transcribed and thus provide, by virtue of an altered chromatin configuration, an opportunity for recombination with foreign DNA. Additional RNA's have been identified in cell line HA12/7 which are initiated in the left terminal sequences of the integrated Ad12 genome and extend into the abutting hamster cell sequence (Schulz et al. 1987). Alterations of cellular chromatin at sites of DNA replication or active transcription may predispose these sites for genetic exchange. By using adenovirus DNA integration as an experimental model, the present report provides a number of well-analyzed examples which clearly document this concept.

Fig. 1a-f. Mapping of cellular transcripts with homologies to hamster, mouse or human DNA at sites of integration or recombination with adenovirus DNA. The structures of integration sites of adenovirus DNA from the Ad2-transformed hamster cell line HE5 (*a*), from the Ad12-induced mouse tumor CBA-12-1-T (*b*), from the Ad12-induced hamster tumor line CLAC1 (*c*), from the Ad12-transformed hamster cell line HA12/7 (*d*), the structure of the site of recombination between Ad12 and human DNA in the SYREC2 DNA molecule (*e*), and the structure of the integration site in the Ad12-induced tumor T1111(2) (*f*) are shown. The map positions of transcripts in cellular DNA are indicated by solid bars below individual maps. When determined, *arrows* indicate the polarity of transcripts. Symbols: Δ, deletions in viral or cellular DNA; -//-,

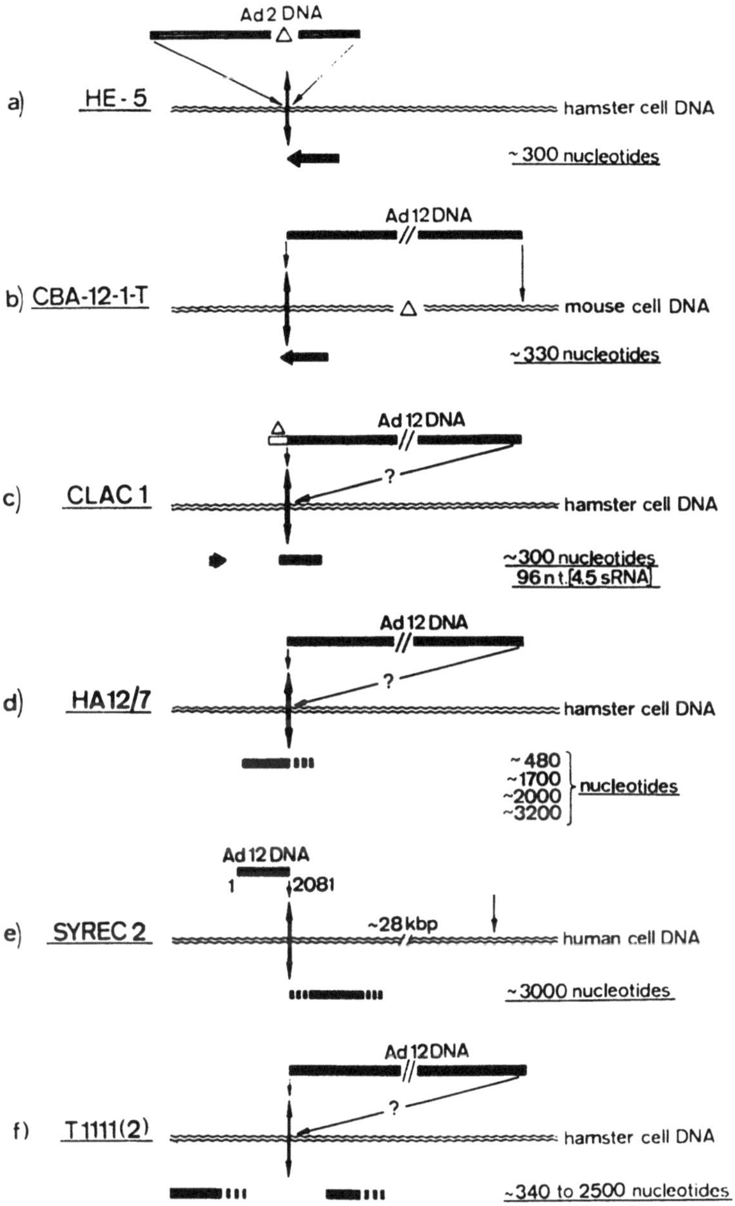

Fig. 1a-f.

interruptions in scale. The approximate lengths of the transcripts are given in nucleotides. In *b*, cellular DNA has been deleted at the site of insertion. In *a*, no cellular nucleotides were deleted. In *c* and *d*, it was not known whether cellular nucleotides were deleted. In *a*, *b* and *d*, terminal deletions of viral nucleotides were < 11 nucleotides (*a*, *b*) or were nonexistent (*d*) and were therefore not indicated. In *c*, 174 nucleotides were deleted (Δ) at the left junction site. In *e*, the left terminal 2081 nucleotides of Ad12 DNA were linked to human cellular DNA. The cellular transcript of about 3000 nucleotides was not precisely mapped (indicated by *broken bar*). In *f* 64 nucleotides of Ad12 DNA were deleted at the left site of junction. The RNA's were not precisely mapped (indicated by *broken bars*)

Insertion and Loss of Viral Genome

Insertion of Ad12 DNA in the Vicinity of an Intracisternal A Particle (IAP) Genome in Tumor Cells. We have previously presented several examples of Ad12 DNA sequences being unstably integrated in Ad12-transformed hamster cells (line T637) (Groneberg et al. 1978) or in an Ad12-induced hamster tumor cell line [H1111(1)] (Kuhlmann et al. 1982). In the instance of the Ad12-induced tumor cell line·H1111(1), it has been demonstrated that the complete loss of Ad12 DNA sequences is fully compatible with the oncogenic phenotype of the cells that have lost the viral genomes. Hence, persistence of Ad12 genes cannot be a precondition in all instances for the maintenance of the oncogenic phenotype. We have now analyzed in more detail another site of Ad12 DNA insertion that has proved unstable upon passage of the cells in culture (Lichtenberg et al. 1987).

In the Ad12-induced hamster tumor T1111(2), about ten Ad12 genome equivalents were integrated at different sites. One of the integrated copies proved unstable and was lost from the cellular genome or rearranged upon passage of the cell line H1111(2) which had been established from this tumor. This unstable site of junction between the left terminus of Ad12 DNA and hamster DNA was cloned, sequenced, and analyzed. Similarly, the preinsertion site was cloned from BHK21 hamster cells, sequenced, and analyzed. The junction site showed several interesting peculiarities. At the left terminus of Ad12 DNA, the first 64 nucleotides were deleted. At a distance of 127 nucleotides to the left from this junction site, an internal dispersed fragment of Ad12 DNA comprising nucleotides 1290 to 1361 of the authentic Ad12 DNA sequence was inserted into cellular DNA in an inverted orientation, relative to the complete Ad12 genome that was located in its vicinity. The 127 nucleotide sequence between the intact Ad12 genome and the separate 72 bp Ad12 DNA fragment was cellular, but it was not identical to the preinsertion sequence at this location. The sequences flanking the termini of the dispersed 72 bp Ad12 DNA fragment were characterized by direct repeats of nine (ten) nucleotides. To the left of Ad12 nucleotide 1361 in the dispersed 72 bp Ad12 DNA fragment, about 620 cellular nucleotides followed which were identical at the occupied and at the preinsertion sites. It was conceivable that the separate 72 bp Ad12 DNA fragment and the cellular sequence of 127 bp to its right had been transposed en bloc from another unknown, more distant location. Abutting the 620 nucleotides of cellular DNA to the left of this block, the 3' terminal sequence of an endogenous, intracisternal A particle (IAP) genome of hamster cells (Ono et al. 1985) was detected. Except for the deletion of one 23-bp repeat in the long terminal repeat (LTR) of the IAP sequence close to the junction site, the LTR sequence was nearly identical to that in the authentic IAP sequence. Details of these sequence arrangements were schematically shown in Fig. 2. The possible functional significance of the close vicinity of an IAP sequence to an inserted Ad12 genome with respect to the transformation event, to the instability of this site, and the transcriptional activity of this region was not readily apparent. The 620 bp of cellular DNA between the 72 bp Ad12 DNA fragment and the end of the LTR of the hamster IAP sequence were apparently of the unique type. The preinsertion sequence and its environment were also investigated for transcriptional activities. There was no such activity in the approximate region between nucleotides -620 (to the left) and +350 (to the right) relative to the site of Ad12 DNA insertion, but there was transcription outside these boundaries (Fig. 1f).

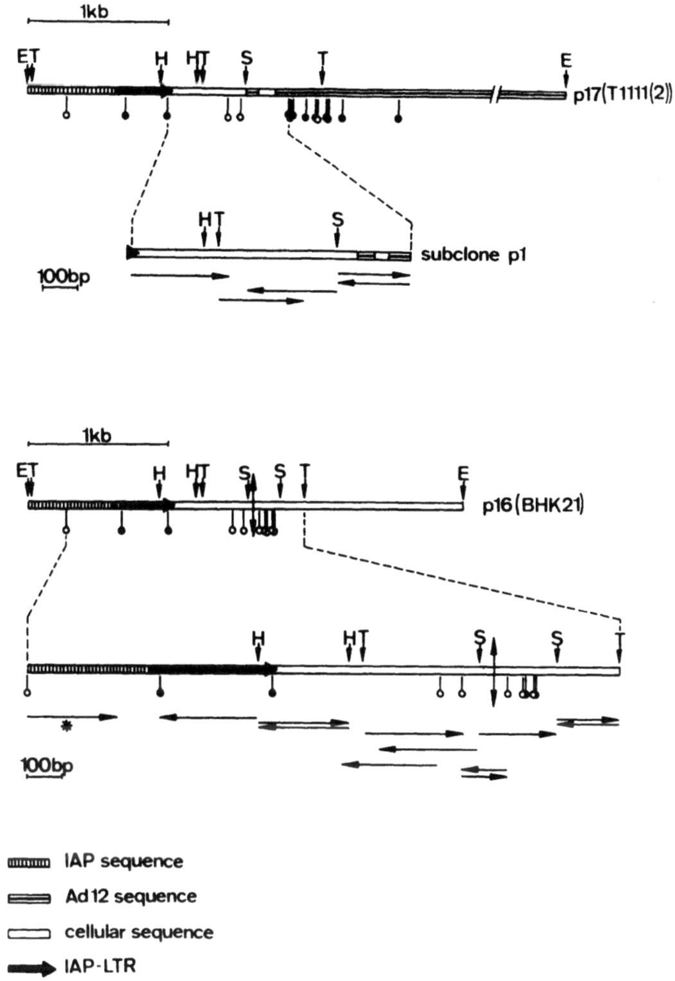

Fig. 2. Mapping of the cloned junction fragment between the left terminus of Ad12 DNA and hamster DNA from tumor T1111(2). One of the junction fragments between Ad12 DNA and cellular DNA was cloned in λgtWES·λB DNA (λ17) or in pBR322 DNA (p17). Detailed maps of the cloned junction site (clone p17 in pBR322 DNA) from T1111(2) DNA, of the p1 HhaI subclone, and of the preinsertion clone (p16 in pBR322 DNA) from BHK21 hamster DNA. Individual restriction sites were designated as indicated. Restriction sites: EcoRI (*E*), HindIII (*H*), SphI (*S*), TaqI (*T*), HaeIII (ᵹ), HhaI (♦). Hamster cell DNA, Ad12 DNA, intracisternal A particle (IAP) DNA, and the IAP long terminal repeat (LTR) sequences were designated by symbols as shown. The horizontal arrows correspond to subfragments that were used in determining the nucleotide sequences of the site of junction and of the preinsertion site in hamster DNA (not shown here; cf. Lichtenberg et al. 1987). The site of Ad12 DNA insertion was marked by a *vertical, double-headed arrow* in the hamster preinsertion sequence

The recombination site analyzed in this report had been chosen because of its instability. In the tumor T1111(2), one or very few copies of Ad12 DNA were integrated at this unstable site. In passage 3 of the hamster tumor cell line H1111(2) this Ad12 DNA molecule was lost from this site or was rearranged. In view of the instability of this insertion site, it might be significant that an endogenous IAP sequence (Ono et al. 1985) lay in close vicinity. It had been noted that this IAP sequence has transposon-like properties (Canaani et al. 1983, Kuff et al. 1983). It would, therefore, not be surprising to find a certain propensity for DNA rearrangement, excision or transposition events in the surroundings of this IAP sequence. It remained a matter of conjecture whether the insertion of Ad12 DNA at this particular site might have been influenced by the presence of the endogenous IAP sequence in its vicinity.

Recombination in a Cell Free System Between Ad12 DNA and the p7 Hamster Cell DNA from a Cloned Preinsertion Site

Detailed studies on the molecular structure of adenovirus DNA integration sites in a number of mammalian species have revealed interesting information on some of the parameters important in foreign DNA insertion. By this approach of molecular archeology, in which the products of recombination events have been studied, which anteceded the time of analysis by many cell generations, only limited insight can be gained into the enzymatic mechanism(s) of foreign DNA insertion into the mammalian genome. We have, therefore, devised alternate ways of investigating genetic interactions between the mammalien host genome and foreign DNA, and have started to work on this type of recombination in a cell-free system. The first results of these experiments will be reported here.

In the Ad12-induced hamster tumor line CLAC1, 10 to 13 copies of Ad12 DNA are integrated into the cellular genome. One site of junction between the left end of Ad12 DNA and hamster DNA has been cloned and sequenced. This analysis has revealed a deletion of the 174 left terminal nucleotides of Ad12 DNA and linkage of repetitive hamster DNA to nucleotide 175 in Ad12 DNA. Between the deleted viral nucleotide sequence and the cellular sequence replacing it, an interesting array of patchy homologies was detected (Stabel and Doerfler 1982). The rationale for the series of experiments described here was to mimic recombination events similar to those that led to the integration of Ad12 DNA in the CLAC1 tumor line.

A cell-free system was developed using nuclear extracts from BHK21 cells to study the recombination between Ad12 DNA and the preinsertion sequence p7 from BHK21 cells. This preinsertion sequence corresponded to one of the sites of Ad12 DNA integration in the tumor cell line CLAC1. This preinsertion sequence had been cloned in the BamHI site in pBR322 DNA (clone p7). Since the p7 hamster DNA sequence had recombined with Ad12 DNA at least once when cell line CLAC1 was generated, this sequence might carry signals not apparent even to computer-aided analyses, but might be recognizable again by cellular and/or viral proteins involved in a similar recombination event. Details of the preparations of nuclear extracts from BHK21 cells have been schematically outlined in Table 2. Ad12 DNA and the linearized, EcoRI-cut p7 plasmid DNA were incubated with prepurified extracts. The plasmid DNA was then recloned in recA$^-$ strains of *E. coli*, and Ad12 DNA-positive colonies were identified by hybridization to [^{32}P]-labeled Ad12 DNA using the method of Grunstein and Hogness (1975). So far, several recombinants have been identified and analyzed in more detail. The schemes in Fig. 3 describe the molecular structure of one of these recombinant plasmids, pR1. Plasmid pR2 (not shown) consists predominantly of Ad12 DNA. In pR1 the restriction analysis revealed that both Ad12 viral and p7 cellular DNA sequences have been incorporated

Table 2. An in vitro system to study recombination between Ad12 DNA and hamster DNA

BHK21 cells

Isolated nuclei
Sonic treatment of nuclei
DEAE-cellulose flow-through
$(NH_4)_2SO_4$ precipitate, dialysis: extract

Experimental design

Ad12 DNA + p7 DNA + extract
Incubate at 37°C   Reextract DNA
Transfect into recA$^-$ strain of *E. coli* (ampicillin selection)
Select Ad12 DNA positive colonies
Restriction and sequence analysis of recombinants

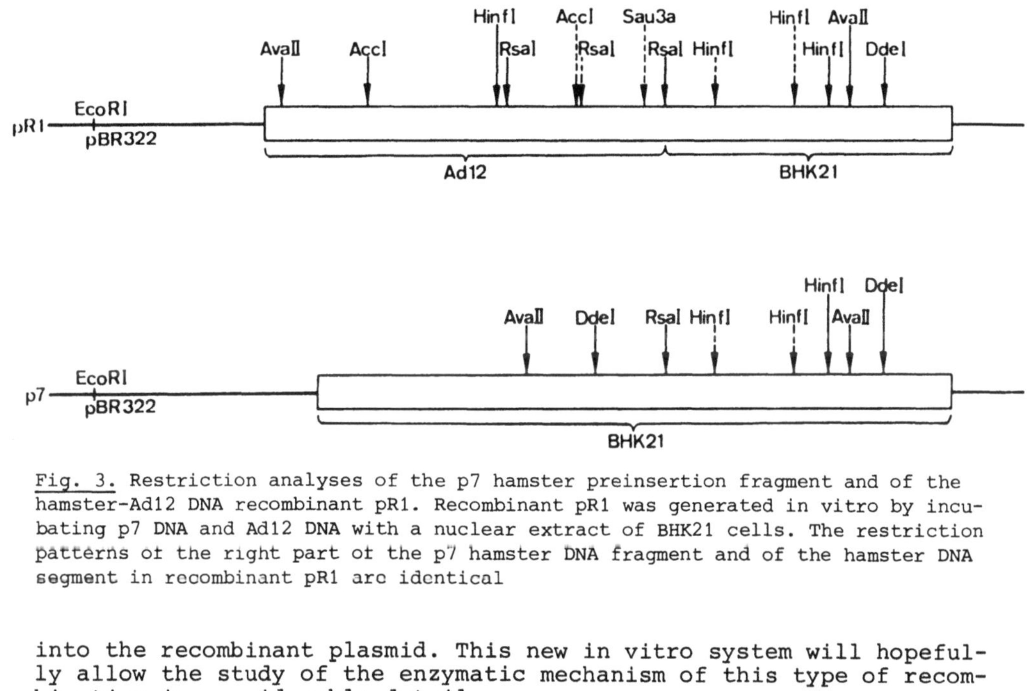

Fig. 3. Restriction analyses of the p7 hamster preinsertion fragment and of the hamster-Ad12 DNA recombinant pR1. Recombinant pR1 was generated in vitro by incubating p7 DNA and Ad12 DNA with a nuclear extract of BHK21 cells. The restriction patterns of the right part of the p7 hamster DNA fragment and of the hamster DNA segment in recombinant pR1 are identical

into the recombinant plasmid. This new in vitro system will hopefully allow the study of the enzymatic mechanism of this type of recombination in considerable detail.

Adenovirus DNA Integration in Productively Infected Cells: Specific Chromosomal Association of Ad12 DNA in Human Cells

In earlier work, we have adduced evidence that adenovirus DNA can also become covalently linked to cellular DNA in productively infected cells (Burger and Doerfler 1974, Schick et al. 1976, Baczko et al. 1978, Neumann and Doerfler 1981, Deuring et al. 1981b, Deuring and Doerfler 1983, Vogel et al. 1986). Recently, we have demonstrated that at early times (2-4 h) after the productive infection of human cells with Ad12, viral DNA becomes associated with human chromosomes. This association

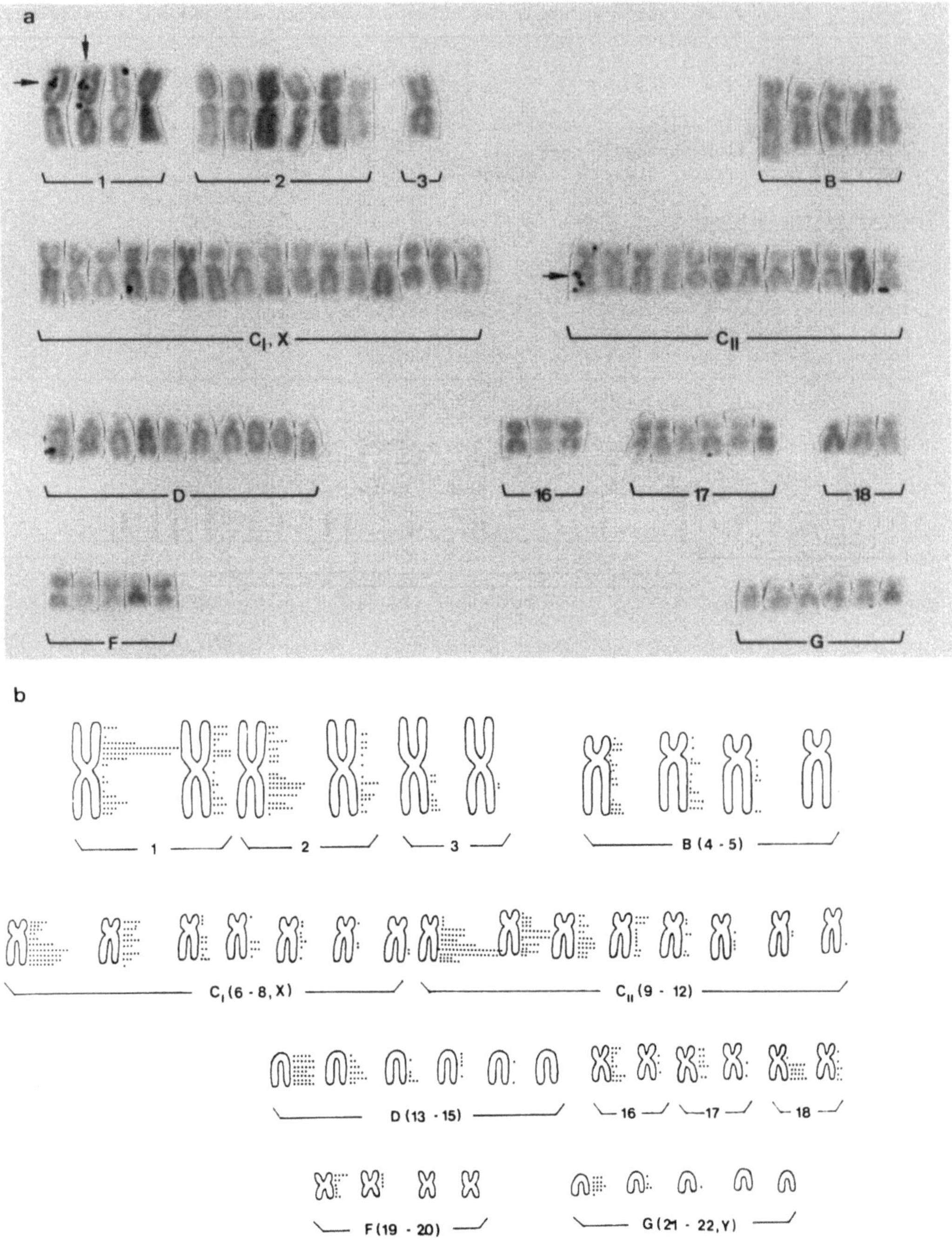

Fig. 4a,b. In situ hybridization of [^{3}H]-labeled Ad12 DNA to chromosomes isolated from human KB cells 2 h after the productive infection with Ad12. *a* Autoradiogram of 77 chromosomes from Ad12-infected KB cells. *b* Distribution of silver grains on individual chromosomes from Ad12-infected KB cells. The autoradiograms were prepared after hybridization in situ to [^{3}H]-labeled Ad12 DNA. Details have been described elsewhere (Vogel et al. 1986). This graph is based on the analysis of 25 metaphases is predominantly, probably not exclusively, with chromosomes of the A and C_{II} groups (Vogel et al. 1986). The data presented in Fig. 4 show that chromosome number 1 in human KB cells can predominantly associate with Ad12 DNA. This finding raises tantalizing questions as to the biological significance of this integration event. Since transcriptionally active sites in the mammalian chromosome have been shown to predispose for the insertion of adenovirus DNA (Schulz et al. 1987), it is conceivable that the integration of adenovirus DNA, perhaps at selective sites, bestows advantages on integrated adenovirus genes in the competition for active transcriptional signals and factors. These advantages may be decisive early in infection in an environment with a high degree of competition for available transcription factors. Currently we are performing experiments that have been designed to test the following working hypothesis: integration of adenovirus DNA or of viral DNA fragments at very early times, i.e., the first few hours, after productive infection of human cells facilitates the efficient expression of very early viral genes whose products are required to activate the transcriptional cascades of adenovirus DNA expression.

Prospects for Further Studies on Viral DNA Integration

With the development of cell-free systems in mammalian cells to study mechanistic aspects of foreign DNA recombination/integration, this field of investigations will open many new vistas. Aside from these biochemical approaches, other fundamental problems deserve intense studies. What is the biological and the evolutionary significance of the apparently high frequency of recombination/integration of foreign DNA with the host genome in mammalian cells? What role can these events play in oncogenesis, in evolution? Possibly the uptake and fixation of foreign DNA into the cellular genome might have played an important role during evolution. What is the immediate advantage or perhaps disadvantage of such mechanisms for the host cell? Can the host cell muster defenses against the intrusion of foreign genetic material into its own genome? Is the well-documented excision event part of these defensive arsenals? As is frequently the case in research on an interesting, generally relevant topic, more questions can be raised than it will be possible to answer without intensive, long-term work.

Since fixation of viral DNA and its excision are important topics in the understanding of viral persistence and its possible cure, these topics will continue to occupy our laboratory's interest, even though we may have to spend another decade or more to make a dent on this highly complex problem. Are there possibilities to strengthen the host cells' defense mechanisms against the uptake of foreign DNA into its own genome or to further the excision of the alien intruder? Perhaps the solution to some of the most pressing problems in viral or retroviral persistence could receive new impetus from this unexpected direction.

Acknowledgments. We thank Petra Böhm for excellent editorial work. Over many years, this research was supported by the Deutsche Forschungsgemeinschaft through SFB74-C1, by the Federal Ministry of Research and Technology, Bonn-Bad Godesberg, Germany (BCT 0390/2), and by the Ministry of Science and Research of the State of Nordrhein-Westfalen (IV B5 - 500 027 86).

References

Baczko K, Neumann R, Doerfler W (1978) Virology 85:557-567
Burger H, Doerfler W (1983) J Virol 13:975-992
Canaani E, Dreazen O, Klar A, Rechavi G, Ram D, Cohen JB, Givol D (1983) Proc Natl Acad Sci USA 80:7118-7122

Deuring R, Doerfler W (1983) Gene 26:283-289
Deuring R, Winterhoff U, Tamanoi F, Stabel S, Doerfler W (1981a) Nature (London) 293:82-84
Deuring R, Klotz G, Doerfler W (1981b) Proc Natl Acad Sci USA 78:3142-3146
Doerfler W (1968) Proc Natl Acad Sci USA 60:636-643
Doerfler W (1970) J Virol 6:652-666
Doerfler W (1982) Curr Top Microbiol Immunol 101:127-194
Doerfler W, Gahlmann R, Stabel S, Deuring R, Lichtenberg U, Schulz M, Eick D, Leisten R (1983) Curr Top Microbiol Immunol 109:193-228
Doerfler W, Gahlmann R, Stabel S, Deuring R, Schulz M, Lichtenberg U, Eick D, Jessberger R, Leisten R (1985) In: Becker Y (ed) Recombinant DNA research and viruses. Cloning and expression of viral genes. Martinus Nijhoff Publ., Boston, pp 125-171
Eick D, Doerfler W (1982) J Virol 42:317-321
Gahlmann R, Doerfler W (1983) Nucleic Acids Res 11:7347-7461
Gahlmann R, Leisten R, Vardimon L, Doerfler W (1982) EMBO J 1:1101-1104
Gahlmann R, Schulz M, Doerfler W (1984) EMBO J 3:3263-3269
Groneberg J, Sutter D, Soboll H, Doerfler W (1978) J Gen Virol 40:635-645
Grunstein M, Hogness DS (1975) Proc Natl Acad Sci USA 72:3961-3965
Kuff EL, Feenstra A, Lueders K, Smith L, Hawley R, Hozumi N, Shulman M (1983) Proc Natl Acad Sci USA 80:1992-1996
Kuhlmann I, Doerfler W (1982) Virology 118:169-180
Kuhlmann I, Achten S, Rudolph R, Doerfler W (1982) EMBO J 1:79-86
Lichtenberg U, Zock C, Doerfler W (1987) J Virol 61:2719-2726
Neumann R, Doerfler W (1981) J Virol 37:887-892
Ono M, Toh H, Miyata T, Awaya T (1985) J Virol 55:387-394
Rekosh DMK, Russell WC, Bellett AJD, Robinson AJ (1977) Cell 11:283-295
Robinson AJ, Younghusband HB, Bellett AJD (1973) Virology 56:54-69
Schellner J, Stüber K, Doerfler W (1986) Biochim Biophys Acta 867:114-123
Schick J, Baczko K, Fanning E, Groneberg J, Burger H, Doerfler W (1976) Proc Natl Acad Sci USA 73:1043-1047
Schulz M, Doerfler W (1984) Nucleic Acids Res 12:4959-4976
Schulz M, Freisem-Rabien U, Jessberger R, Doerfler W (1987) J Virol 61:344-353
Southern EM (1975) J Mol Biol 98:503-517
Stabel S, Doerfler W (1982) Nucleic Acids Res 10:8007-8o23
Stabel S, Doerfler W, Friis RR (1980) J Virol 36:22-40
Starzinski-Powitz A, Schulz M, Doerfler W (1982) EMBO J 1:493-497
Vogel S, Rosahl T, Doerfler W (1986) Virology 152:159-170

Disorders of Cellular Neuro-Functions by Persistent Viral Infections

K. Koschel[1]

Introduction

In the central nervous system (CNS) cytocidal viruses can destroy cells by cytopathic effects. In some cases, however, cytocidal viruses can establish persistent infections without direct cell destruction and without influence on cell growth and survival.

We are interested in virus persistence in cells of the CNS which can be the cause of heavy progressive CNS diseases developing slowly over periods of months or years. Such infections mostly lead to a fatal end. Pathophysiology and development of these kinds of virus diseases are complex processes involving viral influences and several immunological responses of the host.

In the CNS cooperations between an immense number of specialized cells is organized. This cooperation is based on:

1. Biosynthesis, storage, release and re-uptake of neurotransmitters and neurohormones.
2. The recognition of such signal molecules by specific membrane receptors followed by an inactivation of the signal.
3. The generation of chemical signals called second messengers inside the cells.
4. The generation of electrical potentials based on chemically regulated and voltage dependent ion channels.
5. Specific ion transporters in astrocytes involved in the regulation of K^+ ion homeostasis in the surroundings of neurons.

One has to expect that impairments of these specialized functions without cell destruction will cause disorders in the brain. It would be very difficult to analyze in vivo such specific impairments caused by virus persistence in specific brain cells. Further, immunological processes will influence infected cells and it would also be difficult to discriminate between direct viral influences and immunologically caused events. Therefore to overcome these difficulties we used characterized CNS-derived cell lines with specialized neuro-functions as models in vitro and established persistent viral infections in the cells. We have studied some neural functions in these persistent infected cells and compared them with the functions of uninfected cells of the same type. Very helpful to us was the availability of many neural cell lines. In this respect we have used the rat astrocytoma C6 cell line (1) persistently infected with the paramyxo viruses measles-SSPE (C6/SSPE) and canine distemper (C6/CDV) and the mouse neuroblastoma x rat glioma hybrid cell line 108CC15 created by Amano and Hamprecht (2), an often used cell line for basic studies on neural functions. We infected this cell persistently by rabies virus. Another interesting cell line used in many laboratories as model for sympathetic noradrenergic neuron cells is the rat pheochromocytoma PC12 cell of Lloyd Greene and co-

[1]Institut für Virologie und Immunbiologie, Versbacher Str. 7, D-8700 Würzburg, FRG

38. Colloquium Mosbach 1987
Molecular Basis of Viral and Microbial Pathogenesis

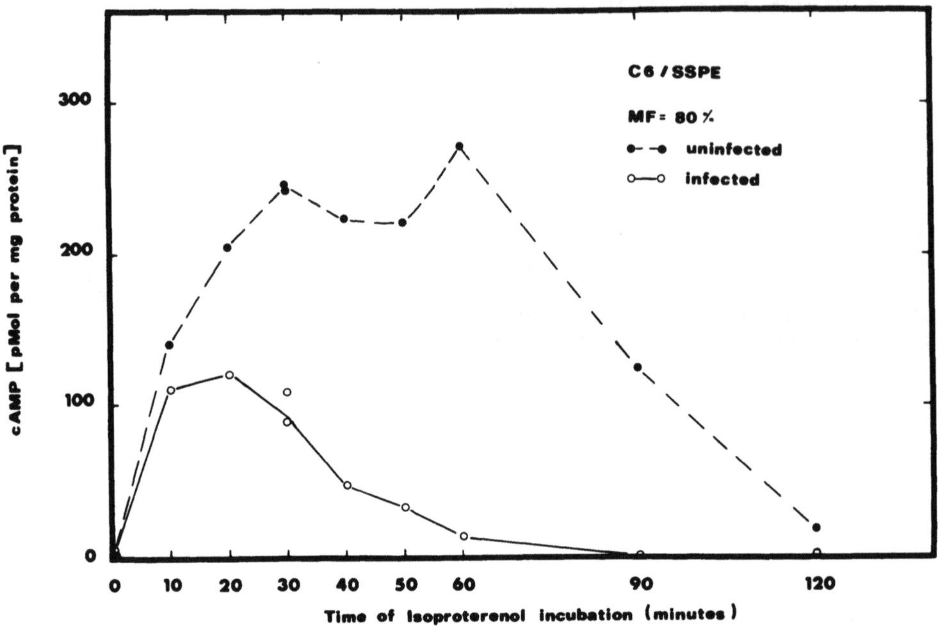

Fig. 1. Intracellular cAMP concentration after addition of D,L-isoproterenol (10^{-4} M) to C6 cells (●——●) and C6/SSPE cells (o——o). 80% of C6/SSPE cells show strong immunofluorescence for viral antigens

workers (3) after differentiation by nerve growth factor (NGF). At present we are working with this cell line. Unfortunately it was not possible to infect the different cell lines persistently with the same type of virus. This chapter shows some results on C6 cells, C6/SSPE cells, C6/CDV cells, 108CC15/rabies cells and their uninfected counterparts.

The viruses used have some proteins associated with the genomic RNA, one protein called M-protein, associated with the envelope membrane and also with the nucleo-capsid structure, and further two typical integral membrane proteins inserted into the virus phospholipid bilayer envelope. Cells persistently infected by these viruses normally show the viral membrane proteins inserted into the cell plasma membrane and the other viral proteins accumulated inside the cell in the cytoplasm. In our C6/CDV system, however, viral proteins are exclusively present in the cytoplasm. The distribution of viral proteins in the infected cells can be demonstrated by common immunofluorescence techniques.

Receptors Coupled to Adenylate Cyclase

Two basic types of receptors are coupled to adenylate cyclase (Fig. 3a):

1. Receptors stimulating the enzyme via a stimulating regulatory GTP binding protein to produce 3'5'-cAMP after binding of a receptor agonist molecule.
2. Receptors inhibiting this stimulation after binding of an agonist via an inhibiting regulatory GTP binding protein.

We investigated our persistently infected cells for the function of such receptor systems and found strong impairments compared with uninfected cells.

a) The β-Adrenergic Receptor System in Persistently Measles SSPE or CDV Infected Rat Glioma Cells. In C6/SSPE and C6/CDV cells we found the following results (4,5):

1. The overall cAMP response after addition of catecholamines is strongly reduced up to 80% in infected cells (Fig. 1).
2. Binding of ^{3}H-dehydroalprenolol to membrane preparations of C6 or C6/SSPE cells was identical.
3. Binding studies in C6/CDV cells show that cells have lost 50% of their β-receptors. Remaining receptors did not change their binding affinity.
4. Both persistently infected cell systems are defect in the coupling of the regulatory GTP binding protein to the adenylate cyclase. Coupling was assayed in membrane preparations in vitro by fluoride, GTP, or Gpp NHp which normally stimulate the adenylate cyclase via the GTP binding protein, but cannot stimulate in membranes of infected cells.
5. Incubation of membranes in presence of forskolin which activate the enzyme directly gives the conclusion that the enzyme unit itself has a normal activity.

The data are summarized in Fig. 2.

b) Prostaglandin and Opiate Receptors in 108CC15/Rabies Cells. The adenylate cyclase system can be regulated by stimulating and inhibiting receptor systems. Some data are given here for the stimulating PGE_1-receptors and inhibiting opiate receptors (δ-type) of rabies infected 108CC15 cells (6-8):

1. A strong inhibition of prostaglandin E_1 or prostaglandin $F_{2\alpha}$-stimulated cAMP synthesis is observed in infected cells.
2. Infected cells show a total loss of the inhibitory opiate receptor function because opiates do not inhibit the remaining PGE_1 induced cAMP response.
3. In vitro incubations of membrane fractions from infected and uninfected cells in presence of GTP, GppNHp or fluoride do not stimulate the adenylate cyclase.
4. The adenylate cyclase unit itself is active as normal. This was shown by membrane incubations in the presence of forskolin. Membranes from infected cells have higher adenylate cyclase activities than control cells because the function of inhibiting regulatory protein is lost as will be shown.

We analyzed the defects of the opiate receptor system more in detail:

1. Binding of ^{3}H-etorphin, an agonist, to membrane fractions show a drastic decrease of affinity in infected cells, but the number of receptors was unchanged.
2. Persistently infected cells are unable to couple opiate receptors to the inhibiting regulatory GTP binding proteins. This can be measured by the opiate-stimulated GTPase activity in this unit in vitro, which is not stimulated in membranes of infected cells.

A scheme will summarize the results (Fig. 3b).

Sodium Channels

Different types of Na^+-channels with different pharmacological properties are described in literature. These properties can be used to specify the channels. Voltage-dependent Na^+ channels are opened by the al-

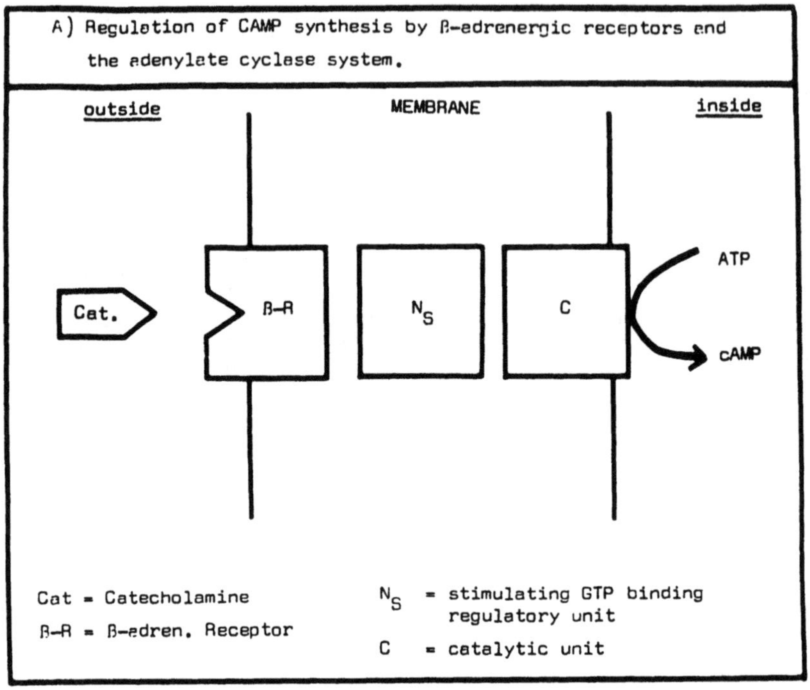

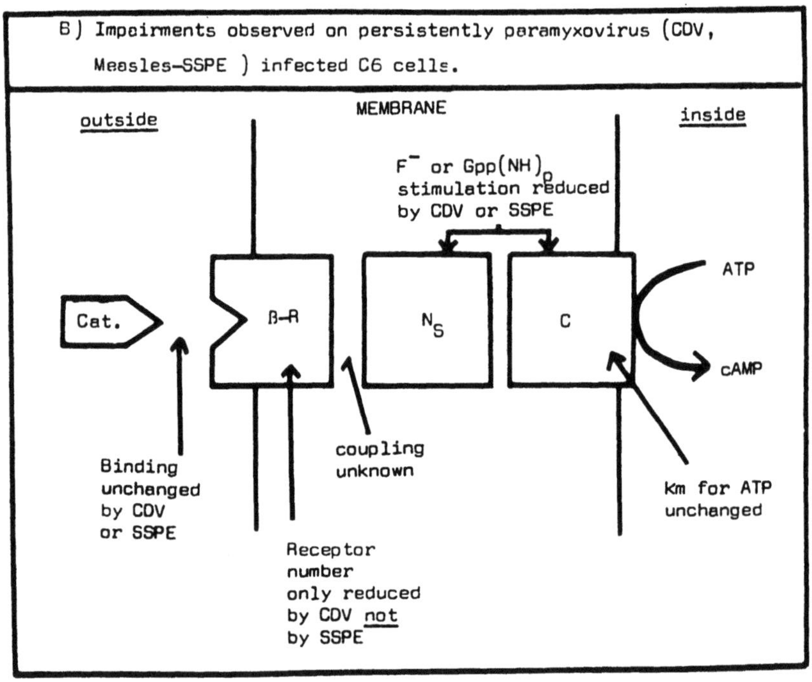

Fig. 2A,B. Scheme of the β-adrenergic receptor/adenylate cyclase system *(upper part)* and impairments of C6/SSPE and C6/CDV cells indicated in the same scheme *(lower part)*

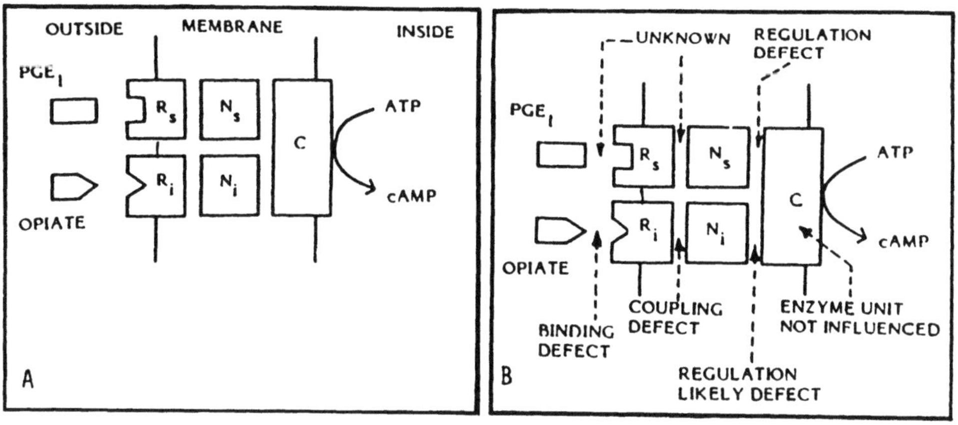

Fig. 3A,B. A scheme of PGE_1 receptor/adenylate cyclase system, counteracted by the opiate receptor system in 108CC15 cells. B Impairments of the system in persistently rabies virus infected 108CC15 cells. R_s PGE_1 receptor; R_i opiate receptor; *C* catalytic cyclase unit; N_s stimulating GTP binding regulatory protein; N_i inhibiting GTP binding regulatory protein

kaloids veratridine or aconitine (9). Low passages of 108CC15 cells have such channels (10). Another type of sodium channels, so-called silent channels, can only be fully opened by the synergistic action of the alkaloids veratridine or aconitine together with alpha-scorpiotoxin (11). These channels are present in C6 rat glioma cells, in primary astrocytes of rat and in some other cell types. Voltage-dependent and silent Na^+ channels are closed by tetrodotoxin. From practical points of view we measure the channel transport with ^{14}C-guanidinium ions instead of radioactive $^{22}Na^+$-ions. I will now present data on silent Na^+ channels of C6 cells, C6/CDV and C6/SSPE cells.

C6/CDV Cells and C6/SSPE Cells. In C6/CDV cells we do not find changes of silent Na^+ channel function (Fig. 4). However, in C6/SSPE cells a strong inhibition of the alkaloid/scorpiotoxin stimulated ion influx is observed (Fig. 5a). The unstimulated influx is equal (Fig. 5b). Unfortunately, the function of silent Na^+ channels is still unknown. Therefore it is not possible to speculate about the consequences of such impairment.

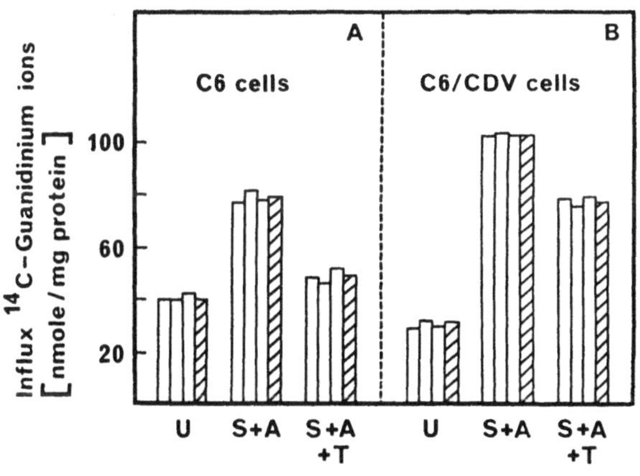

Fig. 4A,B. Na^+ channel influx of ^{14}C-guanidinium ions into uninfected C6 (A) and persistently CDV infected C6/CDV cells (B). *U* spontaneous influx into untreated cells; *S + A* scorpion toxin (10^{-5} M) and aconitine (10^{-4} M) treated cells; *S + A + T* scorpion toxin (10^{-5} M), aconitine (10^{-4} M) and tetrodotoxin (10^{-6} M) treated cells. *Open columns* are single experiments using cells grown in 3-cm Ø Petri dishes. *Hatched columns* show the mean of the single experiments

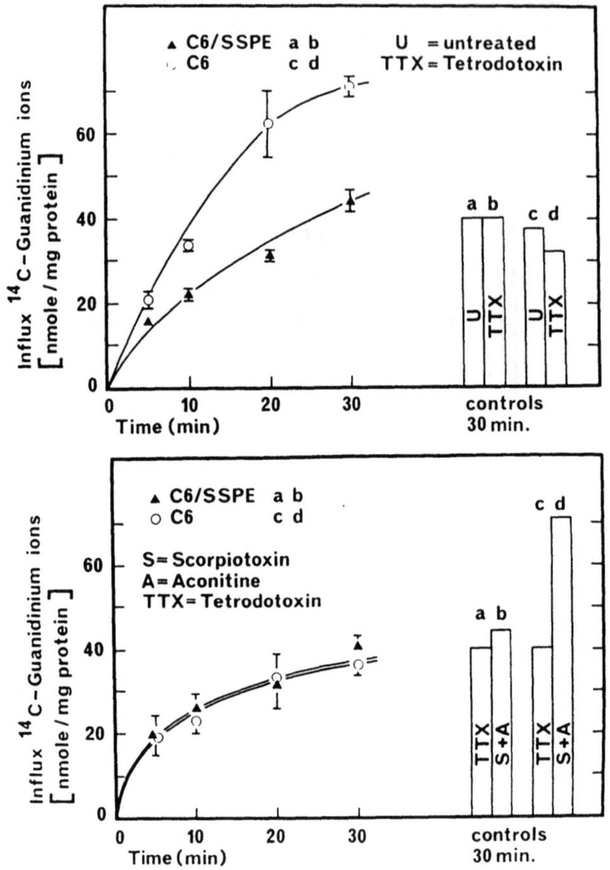

Fig. 5. Na^+ channel influx of ^{14}C-guanidinium ions into uninfected C6 cells (o——o) and persistently measles SSPE virus infected C6/SSPE cells (▲——▲). *Upper diagram* Influx kinetics for the cells treated with scorpion toxin (10^{-5} M) and aconitine (10^{-4} M). *Columns* represent 30 min controls: *U* cells without toxins; *TTX* cells treated with scorpion toxin, aconitine and tetrodotoxin (10^{-6} M) together. *a* and *b* C6/SSPE cells; *c* and *d* C6 cells. *Lower diagram* Influx kinetics as before, but using C6 cells (o——o) and C6/SSPE cells (▲——▲) without any toxin. *Columns* represent 30 min controls: *TTX* cells treated together with scorpion toxin, aconitine and tetrodotoxin; *S + A* cells treated with scorpion toxin and aconitine, *a* and *b* C6/SSPE cells; *c* and *d* C6 cells. All values are means of triplicate determinations

Na^+, K^+, Cl^--Co-Transport and Na^+, K^+-Pump in C6/SSPE and C6/CDV Cells

Associated with astrocytes, as Dr. Tas show in our laboratory (12), is a transport function most likely involved in the K^+ homeostasis in neuron surrounding after release of these ions by action potentials. This function is a Na^+, K^+, Cl^- co-transporter with the following characteristics:

1. It is only functioning if Na^+, K^+ and Cl^- ions are present together (Fig. 6).
2. Hill plots of the data show a strong coupling of the three ions with a stoichiometry of 1 Na^+, 1 K^+ and 2 Cl^- ions (Fig. 7).
3. The transport depends on the Na^+ gradient formed by the Na^+/K^+ pump.
4. The co-transport is inhibited by kidney loop diuretics such as bumetanide and furosemide (Fig. 8).

Both Na^+/K^+ transport and Na^+, K^+, Cl^- co-transport can be measured by the uptake of $^{86}Rb^+$-ions as substitute for potassium into the cells. Na^+/K^+ pump activity was measured as the ouabain-sensitive part of $^{86}Rb^+$ uptake and co-transport activity as the loop diuretics-sensitive part of $^{86}Rb^+$ uptake. Are there impairments of the co-transport and the Na^+/K^+ pump by persistent virus infections? The results to this question are shown in Fig. 9.

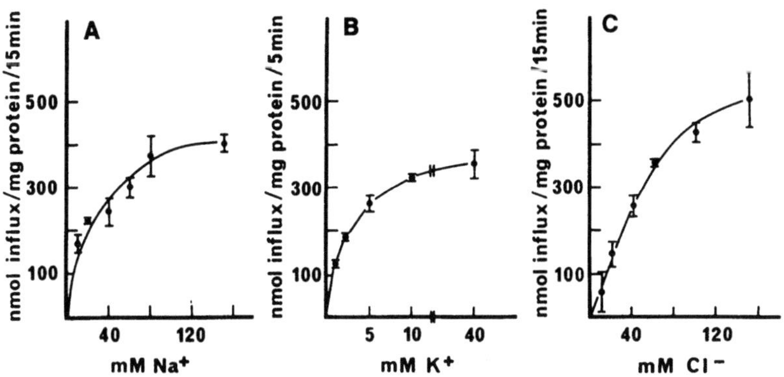

Fig. 6A-C. Dependence of the activity of the $Na^+/K^+/Cl^-$ co-transporter on the extracellular Na^+ (A), K^+ (B) and Cl^- concentrations (C). The medium consisted of 150 mM NaCl, 5 mM KCl, 2 mM $CaCl_2$, 0.4 mM $MgSO_4$, 25 mM glucose and 25 mM Hepes-Tris pH 7.3. Na^+ was replaced by cholinium chloride in (A), K^+ was replaced by Na^+ at concentrations lower than 5 mM and added as K-gluconate at concentrations above 5 mM in (B) and Cl^- respectively was replaced by gluconate in (C). Preincubation was for 15 min in media without Na^+ (A), without K^+ (B) and without Cl^- (C). The incubation was started by the addition of media containing different amounts of resp. Na^+, K^+ or Cl^- and 1 µCi $^{86}Rb^+$/ml. Data are means ± SD of triplicate determinations

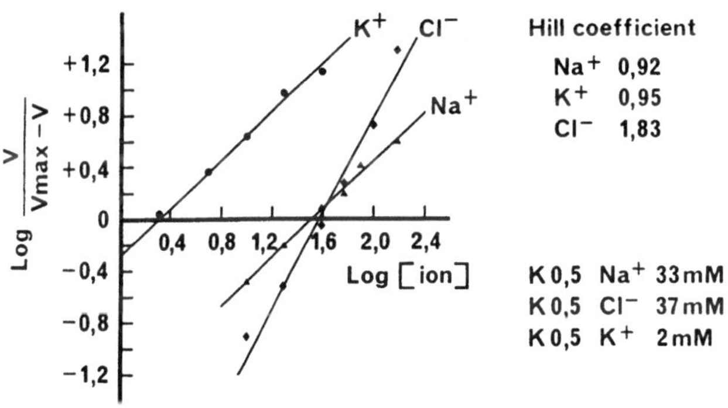

Fig. 7. Hill plots of Na^+, K^+, Cl^- co-transport of rat astrocytes. Data from Fig. 6 were used. The V_{max} values for Na^+, resp. K^+ activated co-transport were determined using Lineweaver-Burk plots of the same data. The V_{max} for the Cl^- activated co-transport was determined using a plot of 1/v against $1/[Cl^-]^2$

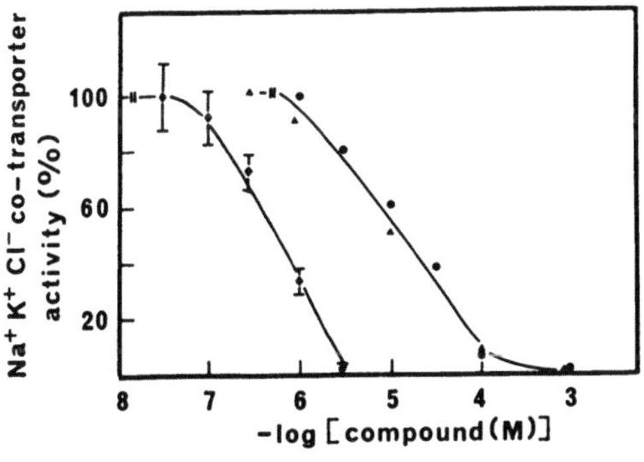

Fig. 8. Effect of bumetanide and furosemide on the activity of the Na^+,K^+,Cl^--co-transporter in primary cultures of astrocytes. Preincubation for 15 min in medium without inhibitors was followed by an incubation of 15 min in medium containing the indicated amount of inhibitor (◆—◆, bumetanide; ●—●, ▲—▲, furosemide) and 1 µCi/ml $^{86}Rb^+$. 100% co-transport activity was taken as the difference between the $^{86}Rb^+$ influx in the absence and presence of 1 mM furosemide. The co-transport activity in the presence of different amount of furosemide and bumetanide was expressed as % of this maximum activity. The bumetanide experiment was performed in duplicate, the furosemide experiment were single determinations of two different experiments

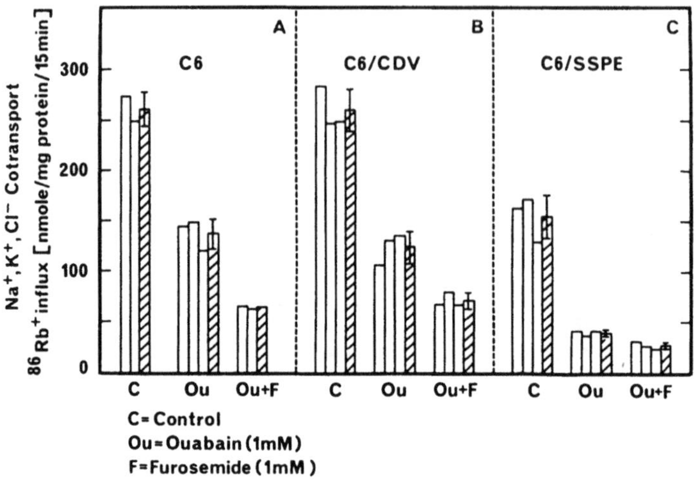

Fig. 9A-C. Na^+, K^+, Cl^- co-transport activity of (A) C6, (B) C6/CDV and (C) C6/SSPE cells. Cells in 3-cm Ø Petri dishes received 1.5 ml preincubation buffer (NaCl 150 mM; KCl 5 mM; $CaCl_2$ 2 mM; $MgSO_4$ 0.4 mM; glucose 25 mM; Hepes-Tris (pH 7.3) 25 mM). After 15 min incubation (37°C) cells were quickly rinsed (3 times) with ice-cold PBS, scraped off in 1 ml 5% tricloroacetic acid and centrifuged. Sediment was used for protein assay. Supernatant was measured for radioactivity by Cerenkov radiation. *C* control; *Ou* Quabain (1 mM); *F* fursosemide (1 mM). *Open columns* are single measurements; *hatched columns* are the means

The Na^+, K^+, Cl^- co-transport is not influenced in the C6/CDV cells, but totally inhibited in C6/SSPE cells. However, in both cases, the activity of the Na^+/K^+ pump is unaltered. A result of the loss of co-transport activity in vivo could be that impairments of K^+ homeostasis could take place. Therefore indirect disturbances of electrophysiological processes could be the consequence.

Antiviral Antibodies and Virus Persistence

This last section presents experiments of effects of antiviral antibodies on persistently infected cells. The experiments should mimic the in vivo situation in the CNS, where oligoclonal antiviral antibodies without complement are observed in the cerebrospinal fluid. Cultivation of C6/SSPE cells in presence of complement-free antiviral antibodies for some passages results in a total loss of viral membrane glycoproteins from the cell surface, and subsequently in a loss of intracytoplasmic viral proteins, as can be shown by immunofluorescence (13, 14). The cells, however, are not cured, since after removal of the antiserum viral antigens re-appear very quickly. Cloning experiments in presence of antibodies lead to the same conclusion. Using Southern blot hybridization, we found that viral mRNA decreases, but viral genomic RNA and a bicistronic intermediate RNA can still be detected after continuous antibody treatment. In vitro translation experiments in a reticulocyte lysate system also show indirectly a strong decrease of viral mRNA because viral proteins are not synthesized.

One can ask what has happened with the impaired functions of C6/SSPE cells like β-receptor response or the silent Na^+ channels after antigenic modulation by antiviral antibodies? The answer is shown for the β-receptor response in Fig. 10. We can see that the β-response is re-

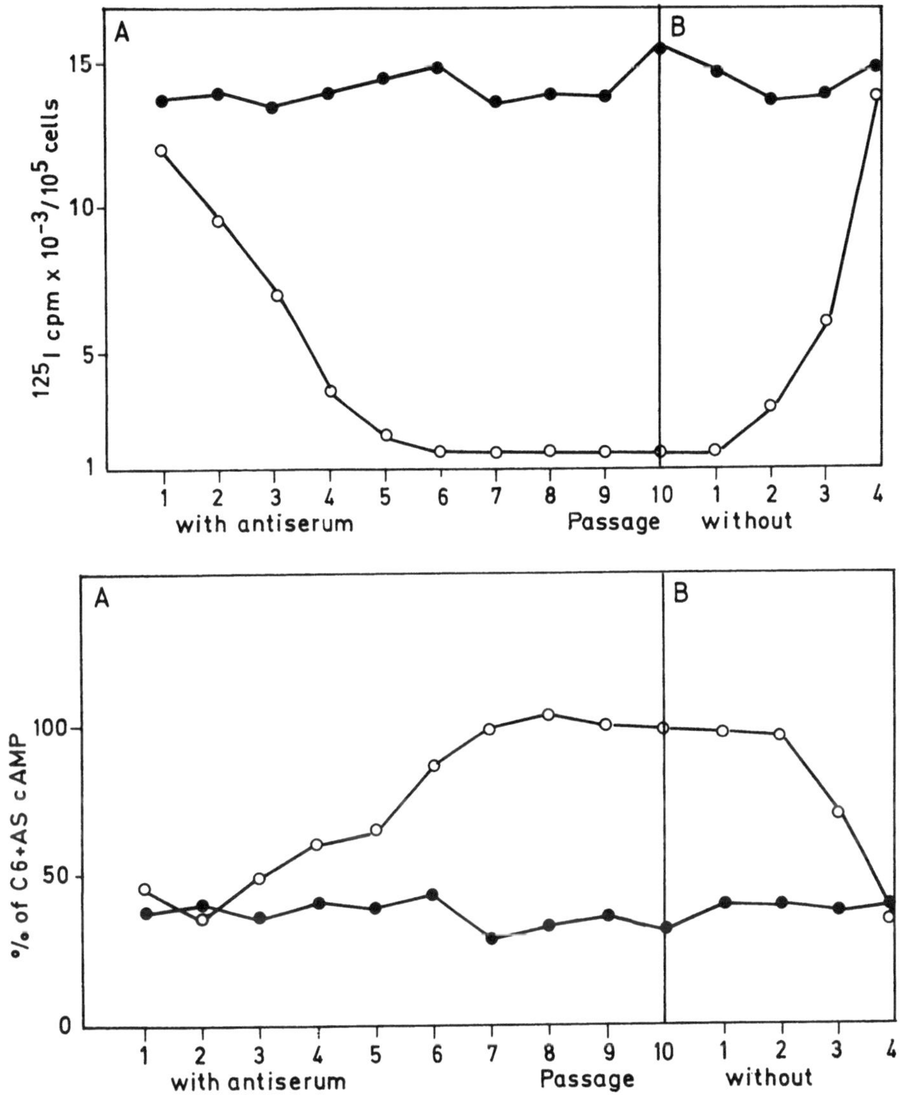

Fig. 10A,B. *Upper part* Virus antigen inserted in the cell membrane was measured by counts bound of ^{125}I-labeled anti-human IgG to cells incubated with human SSPE antiserum. A Amount of virus antigen present after antiserum treatment: (•) C6/SSPE cells grown in antiserum for 1 h only; (o) C6/SSPE cells maintained in antiserum. B Amount of virus antigen present after removal of antiserum: (•) C6/SSPE cells grown in antiserum for 1 h only; (o) C6/SSPE cells released from antiserum treatment. Cells were passaged every 3 - 4 days. *Lower part* Measurement of isoproterenol stimulated cAMP synthesis expressed as a percentage of control levels in uninfected C6 cells grown in SSPE antiserum. A After treatment with antiserum; (•) untreated C6/SSPE cells; (o) C6/SSPE cells maintained in SSPE antiserum. B After removal of antiserum: (•) previously untreated; (o) previously treated with antiserum. Cells were passaged every 3 - 4 days

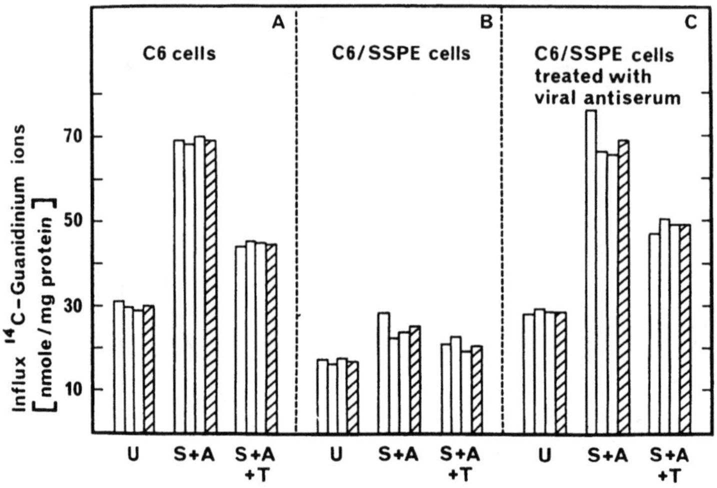

Fig. 11A-C. Na^+ channel influx of ^{14}C-guanidinium ions into (A) C6 cells. (B) C6/SSPE cells and immuno modulated C6/SSPE cells (C), 7 passages treated with 50 HAI units/ml antiviral antiserum. *U* spontaneous influx into cells without toxins; *S + A* scorpion toxin (10^{-5} M) and aconitine (10^{-4} M) treated cells; *S + A + T* scorpion toxin (10^{-5} M), aconitine (10^{-4} M) and tetrodotoxin (10^{-6} M) treated cells. *Open columns* show single experiments using cells grown in 3-cm Ø Petri dishes. *Hatched columns* show the mean of the single experiments

stored if the viral antigens disappear in the course of treatment. After removal of antiviral antibodies the antigens re-appear very quickly and the receptor function is lost again. We observed an analogous situation for the impairment of silent Na^+ channels (Fig. 11).

Even with monoclonal antibodies directed against virus hemagglutinin glycoprotein, we observed the modulation of all viral proteins. Therefore it seems likely that behind the phenomenon a transmembrane signal mechanism is hidden which causes the modulation inside the cells. This, however, is a subject of speculation and will be investigated in future experiments.

Immuno Modulation and Cytotoxic T-Cells. Normally, cells carrying viral antigens on their surface will be eliminated by immunological responses like complement lysis or lysis by cytotoxic T cells which recognize self and nonself antigens together. Complement is not present in the brain. After treatment of infected cells with antiviral antibodies all viral antigens on the cell surface should be removed. Therefore cytotoxic T cell killing should not take place. This is shown on persistently rabies virus-infected and immuno-modulated 108CC15/rabies cells carrying mouse C3H histocompatibility antigens (15). Viral antigen-bearing infected cells are killed by primed cytotoxic spleen lymphocytes from C3H mice (Fig. 12). We used ^{51}chromium release from prelabeled target cells to measure this lysis. As Fig. 12 shows, uninfected and immuno modulated cells are not killed. After removal of anti-rabies antiserum from the culture medium viral, antigens re-appear and cells are killed. Primed T cells from other mouse strains do not kill the infected C3H target cells. Therefore it is a H 2-restricted phenomenon.

These observations of immuno modulation and cytotoxic T cell attack seem to reflect the situation of persistent viral infections in the

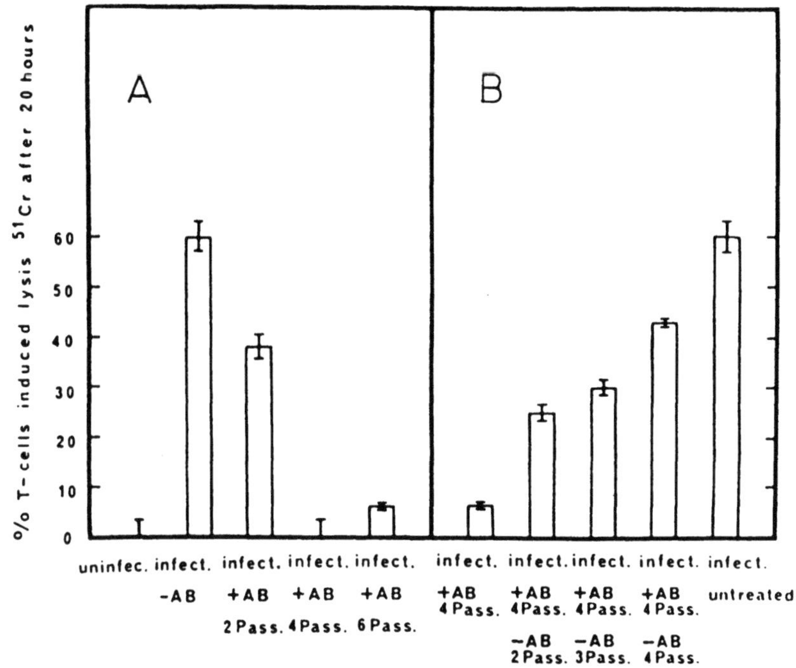

Fig. 12A,B. Antigenic modulation of viral antigens by human antiviral antibodies and cytotoxic T cell lysis of persistently rabies virus infected 108CC15 cells (C3H mouse neuroblastoma x rat glioma). Spleen lymphocytes from rabies virus antigen (purified, Pasteur Institute, Paris) immunized C3H mice were used as effector cells (effector cells : target cells = 50 : 1). Cytotoxic lysis was measured after 1.5 h prelabeling of target cells with ^{51}Cr by release of the radioactivity after 20 h incubation. A Down-modulation and cytotoxic lysis. *AB* antiviral antibodies (50 hemagglutination inhibiting units); pass passages (every second day). B Reversion of antigenic down-modulation after removal of AB and cytotoxic lysis

CNS. The results can explain the establishment and maintenance of persistent infections. Further, the immunological response can represent a pressure of selection for virus mutants or defect viruses which can be isolated from persistently infected cells. Ms. Zinnheimer in our laboratory has shown that treatment of C6/SSPE cells with the monoclonal antibody against virus hemagglutinin leads to down modulation of viral antigens. However, after further passages in *presence* of the antibody, viral antigens re-appear including viral hemagglutinin. The monoclonal antibody, however, does not bind to this modified hemagglutinin molecule because obviously a mutant has been selected.

Conclusion

Data presented here have shown that cell lines derived from the CNS are interesting model systems to elucidate direct influences of persistent viral infections on specialized neural cellular functions. Obviously, viral proteins are involved in the impairments observed. However, at present the special viral proteins and the molecular mechanisms involved are unknown. The observation including immunological responses can be relevant for situations in vivo, but the mechanisms causing immuno modulation are still open questions. We hope to answer some of these problems in the future.

Acknowledgments. This work was supported by the Deutsche Forschungsgemeinschaft, Sonderforschungsbereich 105 (Projekt C 3) and the Wilhelm Sander-Stiftung, Neustadt/ Donau.

References

1. Benda P, Lightbody J, Sato G, Sweet W (1968) Differentiated rat glial strain in tissue culture. Science 161:370-371
2. Hamprecht B (1977) Structural, electrophysiological, biochemical and pharmacological properties of neuroblastoma-glioma cell hybrids in cell culture. Int Rev Cytol 49:99-170
3. Tischler AS, Greene LA (1975) Nerve growth factor-induced process formation by cultured rat pheochromocytoma cell. Nature (London) 258:341-342
4. Halbach M, Koschel K (1979) Impairment of hormone dependent signal transfer by chronic SSPE virus infection. J Gen Virol 42:615-619
5. Koschel K, Münzel P (1980) Persistent paramyxovirus infections and behaviour of β-adrenergic receptors in C6 rat glioma cells. J Gen Virol 47:513-517
6. Koschel K, Halbach M (1979) Rabies virus infection selectively impairs membrane receptor function in neuronal model cells. J Gen Virol 42:627-632
7. Münzel P, Koschel K (1981) Rabies virus decreases agonist binding to opiate receptors of mouse neuroblastoma rat glioma hybrid cells 108CC15. Biochem Biophys Res Commun 101:1241-1250
8. Koschel K, Münzel P (1984) Inhibition of opiate receptor mediated signal transmission by rabies virus in persistently infected mouse neuroblastoma x rat glioma hybrid cells NG-108-15. Proc Natl Acad Sci USA 81:950-954
9. Catteral WA, Nirenberg M (1973) Sodium uptake associated with activation of action potential ionophores of cultured neuroblastoma and muscle cells. Proc Natl Acad Sci USA 81:950-954
10. Reiser G, Scholz F, Hamprecht B (1982) Pharmacological and electrophysiological characterization of lithium ion flux through the action potential sodium channel. J Neurochem 39:228-234
11. Romey G, Jacques J, Schweitz H, Fosset M, Lazdunski M (1979) The sodium channels in non-impulsive cells. Interactions with specific neurotoxins. Biochim Biophys Acta 556:344-353
12. Tas PWL, Massa P, Koschel K (1986) Preliminary characterization of a Na^+, K^+, Cl^--cotransport activity in cultured human astrocytes. Neurosci Lett 70:369-373
13. Barrett PN, Koschel K (1983) Effect of antibody-induced modulation of measles (SSPE) virus membrane proteins on β-adrenergic receptor-mediated adenylate cyclase activity. Virology 127:299-308
14. Barrett PN, Koschel K, Carter M, ter Meulen V (1985) Effect of measles virus antibodies on a measles SSPE virus persistently infected C6 rat-glioma cell line. J Gen Virol 66:1411-1421
15. Koschel K, Münzel P, Halbach M, Metzner R (1986) Persistent rabies virus infection of neuronal cells causes defects in neuroreceptor regulation. In: Levi Montalcini R et al. (ed) Molecular aspects of neurobiology. Springer, Berlin Heidelberg New York Tokyo, pp 158-161

Pathogenetic Aspects of Measles Virus-Persistent Infections in Man

V. ter Meulen[1]

Introduction

Measles virus is a naturally occurring pathogen which is highly contagious for man, causing an acute viral infection in childhood. This disease has been well studied and, with the availability of an effective vaccine, became medically less important. However, the association of measles virus with chronic disorders of the central nervous system and the fact that worldwide measles vaccination programmes are still insufficiently applied has maintained virological interest. In the past, virological studies have been concerned mainly with the analysis of measles virus interaction with cell and host especially in chronic infection of the nervous system and particularly in subacute sclerosing panencephalitis (SSPE) (1). This human slow virus disease is associated with a persistent measles virus infection of brain cells and exhibits epidemiological, clinical, and histopathological features quite different from those seen in acute measles or measles encephalitis. In addition, measles virus has been incriminated as a possible causative agent in multiple sclerosis (MS) based on serological findings and the detection of measles virus RNA in brain tissue by in situ hybridization (2,3). However, despite the major virological discovery that measles virus infections are involved in chronic human disorders, the pathogenetic mechanisms by which such a disease process is induced are still unknown. The mechanisms whereby measles virus persistence is established and maintained in CNS tissue have yet to be defined. Other factors, either host- or virus-derived, must play a pathogenetic role, since the rarity of chronic CNS disorders associated with measles-persistent infections cannot be correlated with the ubiquitous presence of this virus. In the following, the main findings from recent virological investigations of measles virus persistence in brain tissue are summarized in an attempt to discuss and analyze the factors involved in this virus persistence.

Subacute Sclerosing Panencephalitis (SSPE)

Clinical, Neuropathological, and Immunological Findings. SSPE is a rare, slowly progressing disease of the central nervous system (CNS) which occurs primarily in children and young adults (1,5). This disease develops on the basis of a persistent measles virus infection in brain tissue and has been observed in many ethnic groups with an incident of approximately 1 case per 1 million children. History of acute measles is quite often found for the age of 2. The mean age of onset on SSPE is 7 - 10 years of age, with an average interval of 5 to 6 years after acute measles infection. However, cases of SSPE have also been reported at below 2 and over 20 years of age.

[1]Institut für Virologie und Immunbiologie, Universität Würzburg, Versbacher Str. 7, D-8700 Würzburg, FRG

38. Colloquium Mosbach 1987
Molecular Basis of Viral and Microbial Pathogenesis

A clinical course usually follows a stereotype pattern beginning with insidious behavioral changes which, after a period of weeks or months, are superseded by characteristic neurological symptoms. This clinical picture is characterized by various disturbances of motorfunction, myoclonic jerks and epileptic seizures. At a later stage of the disease, a progressive cerebral degeneration with symptoms and signs of decortication occurs. Although, there have been reports of remission, the disease usually follows an ingravescent course over a period of months to years, always ending in death.

Besides this clinical picture, the patients reveal typical electroencephalogram and pathognomonic laboratory findings. Each SSP case tested showed a marked increase in cerebrospinal fluid, IgG, and extremely high measles virus antibody titers in serum and CSF specimens. Measles-specific IgG antibodies represent nearly 10 - 20% of total serum IgG and about 75% of total CSF IgG (6). Moreover, there is always a reduced serum:serum CSF ratio of measles virus antibodies and the finding of oligoclonal CSF IgG which to a large part is measles virus-specific, indicating a local production of measles virus antibodies within the CNS. This strong immune response is directed to all virus structural polypeptides except the matrix protein (M) (7,8).

Neuropathological examinations show a diffuse encephalitis characterized by leptomeningeal lymphocytic infiltrations, perivascular cuffings, consisting of lymphocytes and plasmacells and a diffuse lymphocyte infiltration. The most striking and characteristic morphological changes are the intranuclear inclusion bodies of Cowdry typé A containing measles virus nucleocapsid (10,11,12). These are found in neurons, oligodendroglia cells, and astrocytes. However, a typical feature of measles infection, the formation of giant cells and budding virus particles, has never been seen in SSPE brain sections, suggesting a defect in measles virus replication in CNS tissue. Moreover, no transmission of SSPE, either horizontal or vertical, has been observed, suggesting that the disease is not infectious.

Virological Characteristics. The fact that measles virus structures can be found in any SSPE brain tissue investigated and virus can be occasionally isolated from brain tissue or lymphnodes by co-cultivation methods has identified measles virus as the etiological agent for this disease. Before this chronic CNS disease process develops, the patients have experienced an acute measles virus infection in early childhood and it is therefore likely that a CNS infection by measles virus occurred at this stage, leading to a persistent infection. However, so far no experimental evidence is available which can unequivocally confirm this suggestion. As a result of persistent infection, mutational changes might occur which could lead to the biological and biochemical differences observed between measles virus and isolates from SSPE patients (referred to as SSPE viruses). SSPE isolates vary in their host range, growth rate, plaque size, hemagglutination, and neuropathogenic characteristics, as well as electrophoretic mobility of viral proteins, mRNS's and complexity of viral genomic RNA's (1,13). In addition, differences in antigenicity of certain structural proteins and a reduced sensitivity to neutralization by immune sera in comparison to measles virus have been described. It has also been shown by use of monoclonal antibodies that antigenic differences exist between measles virus isolates from CNS tissue and acute measles which are not distinguishable by polyvalent sera (14,15). These findings demonstrate the occurrence of antigenic changes within the hemagglutinin polypeptide of SSPE virus during the course of a nonproductive persistent infection. However, it should be kept in mind that none of these findings represent a single, stable property of SSPE viruses which is characteristic of measles virus isolates from CNS tissue, since these differences are also observed between different isolates from acute measles virus infections. Therefore, at the present time it seems likely that the virus which has already caused acute measles may also be responsible for the development of SSPE in each patient.

Molecular Biological Findings. The availability of monoclonal antibodies to the structural proteins of measles virus, as well as defined cDNA probes for the different measles virus mRNA's allowed direct evaluation of the state of measles virus persistence in SSPE brain tissue. With the monoclonal antibodies against the nucleocapsid- (N), phospho- (P), hemagglutinin- (H), fusion- (F) and membrane (M) proteins, it could be shown by immunohistological studies that in none of the studied SSPE cases were all five measles virus proteins detectable (16). Only N and P proteins were consistently found in each brain area examined, whereas H, F, and M proteins was only seen occasionally. Within one brain some variation occurred in the expression of measles virus proteins between regions with different degrees of neuropathological changes; however, in the majority of brains studied the same measles virus structural proteins were either present or absent independent of the neuropathological lesions.

The results of the biochemical analysis in these cases were rather complex (17,18). Generally, the mRNA's for measles virus N, P, and M proteins were easily detectable in brain tissues, whereas those for H and F proteins were present but in lower amounts in comparision with the relative concentration of N, P, or M mRNA. Although the absolute amounts varied, the relative amounts of the different virus mRNA's within different parts of the same brain were stable. However, only the mRNA's for the N protein directed synthesis of this polypeptide in in vitro translation systems, whereas P was only seen in 75% of the patients studied. Attempts to detect H or F proteins in in vitro translation products with mRNA isolated from infected brain were unsuccessful, either as a result of the relatively low concentration of these mRNA's or due to functional defects of the transcripts. Of particular interest are the results obtained for M protein. Despite the presence of M mRNA in infected brain tissue, in the majority of cases studied no in vitro product of this M protein was seen. This defect was further analyzed in one case and sequence analysis of the M gene revealed high rates of base transitions and transversions (19). Moreover, one base substitution created a translation stop at codon 12 of the reading frame, indicating that a protein cannot be synthesized.

Pathogenetic Aspects

These recent immunohistological and biochemical investigations demonstrate that measles virus persistence in SSPE brain cells is characterized by restriction of several virus genes. The mRNA's for M, H, and possibly F protein seem to be those most affected. The transcription of M mRNA obviously occurs in most of the patients, but the message is either not translated as a result of accumulation of mutations or if translated the protein synthesized is probably not functional, since ultrastructural changes in infected brain areas do not reveal budding of virus particles. These findings in SSPE-infected brain tissue are very similar to those observed in measles or persistent SSPE virus infections in tissue culture. In some cell lines, M protein is either not translated (20) or is synthesized but subsequently degraded rapidly (21,22), which results in a reduced intracellular pool of this protein. This phenomenon is not unique to measles virus persistence, but may be common in persistent paramyxovirus infection, since it has also been observed in BHK cells persistently infected with Sendai virus (23). However, the nonproductive infection of measles virus does not depend on the M protein alone. Restriction in expression of the measles H gene in SSPE-infected brain tissue, as observed in our cases, could certainly contribute to persistence. This would probably apply to the F protein as well. On the other hand, in SSPE there is apparently no need for the production of infectious virus particles, since the nucleocapsid alone is infectious (24). Therefore, mutations of measles virus genes, except for the N, P, and L genes, are not strictly deleterious for the survival of the persisting virus but may, on the contrary, enable the virus to escape the immune response. For effective elimination of virus-infected cells, the immune effector cells have to recognize the viral structural proteins on the cell surface. In SSPE, infected brain cells do not carry on their cell membrane viral envelope proteins since they are not synthesized. Therefore, infected brain cells are immunologically not destroyed and the replicative complex of measles virus is transported from cell to cell in the CNS via cell processes.

Conclusions

It is surprising that persistent measles virus infection associated with a disease process is detectable only in the Central Nervous System. Obviously, this does not exclude the possibility that this agent also establishes persistence in other tissue, but functional impairment of the organ leading ultimately to clinical symptomatology and morphological changes have not been observed outside the CNS. It is conceivable that the anatomical and physiological peculiarities of the brain play an important role in the extent to which a virus infection can cause dysfunction. In this context it has to be emphasized that the nervous system contains a highly differentiated cell population with complex functionally integrated cell-to-cell connections and highly specialized cytoplasmic membranes. Moreover, CNS tissue is unique in its high metabolic rate and relative lack of regenerative capacity. Consequently, a persistent virus infection might interfere with normal functions, whereas in other organs, with lower energy requirements and higher rates of regeneration, persistent infections could perhaps be tolerated.

The development of life-long immunity with the presence of measurable antibody titers after acute measles strongly suggests persistence of these agents in the host, and this is supported by the observation that multiple re-infection with these agents is not required for this type of immunity. More than 100 years ago, Panum (25) observed on the Faeroe Islands an outbreak of measles after this infectious disease had disappeared for 60 years. Almost every one of the inhabitants acquired measles, except those who were older than 60 years and had contracted measles in early childhood. These persons were protected by an effective immunity in the absence of re-exposure to measles virus. Perhaps measles virus resides in the lymphoid tissue itself and thus produces a continuous immune stimulation, as suggested by the isolation of measles virus from lymphnode material in a case of SSPE. Certainly this aspect has to be further studied and evaluated in relationship to immune persistence.

The ubiquituous occurrence of measles virus and its association with the chronic disease process has led to speculations that this agent may also be responsible for other diseases of the CNS. One of the prime candidates is multiple sclerosis (MS) in which this agent has been incriminated (4). Measles virus infection has been linked to MS on the basis of serological studies. It could be shown by many groups (2) that the majority of MS patients revealed a statistically significant, higher antibody titer than control group. In addition, this increase in measles antibody titers was also demonstrable in CSF associated with an oligoclonal immune response. However, in contrast to SSPE, where the majority of oligoclonal CSF IgG reacts with measles virus, in MS only a very small fraction of oligoclonal IgG carries antimeasles activity. So far, all attempts to isolate virus from MS tissue which could be etiologically linked to the disease have failed and measles virus antigens have not been detected in MS. However, recently, measles RNA has been detected by in situ hybridization in about half the cases of MS, but also in an equivalent proportion of non-MS controls (3,26,27). These observations and other reports of polio and herpes virus genes in human CNS (28) suggest that viral genes may be present in brain quite frequently as a consequence of common viral infections. Such viral genes in brain tissue could either represent an adventitious infection without further consequences or a thread to the host if viral gene expression initiates a pathological process. Certainly, many factors are involved until eventually a disease develops. The complex virus host interplay observed in persistent viral infection of the CNS underlines the necessity for an interdisciplinary approach to this subject.

Acknowledgments. The work described in this manuscript was supported by the Deutsche Forschungsgemeinschaft. I thank Helga Kriesinger for typing this manuscript.

References

1. ter Meulen V, Stephenson JR, Kreth HW (1983) Subacute sclerosing panencephalitis. Compr Virol 18:105-159
2. Norrby E (1978) Viral antibodies in multiple sclerosis. In: Melnich JL (ed) Progress in medical virology. Karger, Basel, pp 1-30
3. Haase AT, Ventura P, Gibbs CJ Jr, Tourtellotte WW (1981) Measles virus nucleotide sequences: Detection by hybridization in situ. Science 212:672-675
4. ter Meulen V, Stephenson JR (1983) The possible role of viral infections in MS and other related demyelinating diseases. In: Hallpike JF, Adams CWM, Tourtellotte WW (eds) Multiple Sclerosis. Chapman & Hall, London, pp 241-274
5. Agnarsdottir G (1977) Subacute sclerosing panencephalitis. In: Waterson AP (ed) Recent advances in clinical virology. Livingstone, Edinburgh, pp 21-49
6. Mehta PD, Kane A, Thormar H (1977) Quantitation of measles virus-specific immunoglobulins in serum, CSF, and brain extract from patients with subacute sclerosing panencephalitis. J Immunol 118:2254-2261
7. Hall WW, Choppin PW (1979) Evidence for the lack of synthesis of the M polypeptide of measles virus in brain cells from SSPE. Virology 99:443-447
8. Stephenson JR, ter Meulen V (1979) Antigenic relationships between measles and canine distemper virus. Comparison of immune response in animals and humans to individual virus-specific polypeptides. Proc Natl Acad Sci USA 76:6601-6605
9. Wechsler SL, Weiner HL, Fields BN (1979) Immune response in subacute sclerosing panencephalitis: reduced antibody response to the matrix protein of measles virus. J Immunol 123:884-889
10. Connolly JH, Allen IV, Hurwitz LJ, Miller JHD (1967) Measles-virus antibody and antigen in subacute sclerosing panencephalitis. Lancet 1:542-544
11. Freeman JM, Magoffin RL, Lennette EH, Herndon RM (1967) Additional evidence of the relations between subacute inclusion-body encephalitis and measles virus. Lancet 2:129-131
12. ter Meulen V, Enders-Ruckle G, Müller D, Joppich G (1969) Immunhistological, microscopical and neurochemical studies on encephalitides. III. Subacute progressive panencephalitis. Virological and immunohistological studies. Acta Neuropathol 12:244-259
13. Fraser KB, Martin SJ (1978) Measles virus and its biology. Academic Press, London New York
14. ter Meulen V, Löffler S, Carter MJ, Stephenson JR (1981) Antigenic characterization of measles and SSPE virus haemagglutinin by monoclonal antibodies. J Gen Virol 57:357-364

15. Carter MJ, Willcocks MM, Löffler S, ter Meulen V (1982) Relationships between monoclonal antibody binding sites on the measles virus haemagglutinin. J Gen Virol 63:113-120
16. Liebert UG, Baczko K, Budka H, ter Meulen V (1986) Restricted expression of measles virus proteins in brains from casess of subacute sclerosing panencephalitis. J Gen Virol 67:2435-2444
17. Baczko K, Carter MJ, Billeter M, ter Meulen V (1984) Measles virus gene expression in subacute sclerosing panencephalitis. Virus Res 1:585-595
18. Baczko K, Liebert UG, Billeter M, Cattaneo R, Budka H, ter Meulen V (1986) Expression of defective measles virus genes in brain tissues of patients with subacute sclerosing panencephalitis. J Virol 59:472-478
19. Cattaneo R, Schmid A, Rebmann G, Baczko K, ter Meulen V, Bellini WJ, Rozenblatt S, Billeter MA (1986) Accumulated measles virus mutations in a case of subacute sclerosing panencephalitis: interrupted matrix protein reading frame and transcription alteration. Virology 154:97-107
20. Carter MJ, Willcocks MM, ter Meulen V (1983) Defective translation of measles virus matrix protein in a subacute sclerosing panencephalitis cell line. Nature (London) 305:153-155
21. Sheppard RD, Raine CS, Bornstein MB, Udem SA (1985) Measles virus matrix protein synthesized in a subacute sclerosing panencephalitis cell line. Science 66:1397-1409
22. Young KKY, Heineke BE, Wechsler SL (1985) M protein instability and lack of H protein processing associated with nonproductive persistent infection of HeLa cells by measles virus. Virology 143:536-545
23. Roux L, Waldvogel FA (1982) Instability of the viral M protein in BHK-21 cells persistently infected with Sendai virus. Cell 28:293-302
24. Rozenblatt S, Koch T, Pinhasi O, Bratasin S (1979) Infective substructures of measles virus from acutely and persistently infected cells. J Virol 32:329-333
25. Panum PL (1847) Iagttageiser anstillede under measlinge-apidemien paa faeröerne i aaret 1846. Bibl Laeger 1:270-344
26. Weiner LP, Johnson RT, Herndon RM (1973) Viral infections and demyelinating diseases. N Engl J Med 288:1103-1110
27. Haase AT, Stowring L, Ventura P, Burks J, Ebers G, Tourtellotte W, Warren K (1985) Detection by hybridization of viral infection of the human central nervous system. Ann NY Acad Sci 436:103-108
28. Brahic M, Haase AT, Cash E (1984) Simultaneous in situ detection of viral RNA and antigens. Proc Natl Acad Sci USA 81:5444-5448

Analysis of the Dual Lineage Specificity of E26 Avian Leukemia Virus

T. Graf[1], S. Grieser[1], and M. Nunn[2]

Defective avian leukemia viruses are a group of retroviruses that carry different oncogenes, cause acute leukemia in chickens and transform hematopoietic cells in vitro (for review, see Graf and Beug 1978, Graf and Stehelin 1982). While most strains induce essentially only one type of hematopoietic neoplasm, the E26 virus constitutes a notable exception in that it causes a leukemia consisting predominantly of erythroblasts and to a lesser extent of immature myelomonocytic cells ("myeloblasts") (Sotirov 1981, Radke et al. 1984). In vitro, the virus efficiently transforms both myeloblasts and erythroblasts depending on the culture conditions (Graf et al. 1979, Radke et al. 1982). Interestingly, E26 is also one of the rare examples of a retrovirus that has acquired two oncogenes during evolution, namely v-*myb* and v-*ets* (For review see Nunn et al. 1987). These two oncogenes, together with viral *gag* sequences, are translated into a nuclear protein of 135 kDa MW (Beug et al. 1982).

In an effort to understand the contribution of v-*myb* and v-*ets* in the transforming ability of and leukemogenicity by E26 virus, we have studied the properties of a mutant of E26 virus in which most of the v-*ets* gene had been deleted (Nunn et al. 1987). The genome structure of this mutant, designated Δ4, is schematically illustrated in Fig. 1 in comparison to the genome of wild type (*wt*)E26. Cells infected by this mutant express a protein of a MW of about 80 kDa, as shown by immunoprecipitation with antibodies to *myb* determinants (unpublished results). To obtain a stock of infectious virus, chick embryo fibroblasts were cotransfected with Δ4E26 DNA and MAV-1 helper virus DNA. The cells were then co-cultivated with bone marrow cells in the presence of crude chicken myelomonocytic growth factor (cMGF; Leutz et al. 1984). After 7 - 10 days in culture, rapidly proliferating cells with a blast-like morphology grew out and the supernatants from those cultures were used as a virus stock.

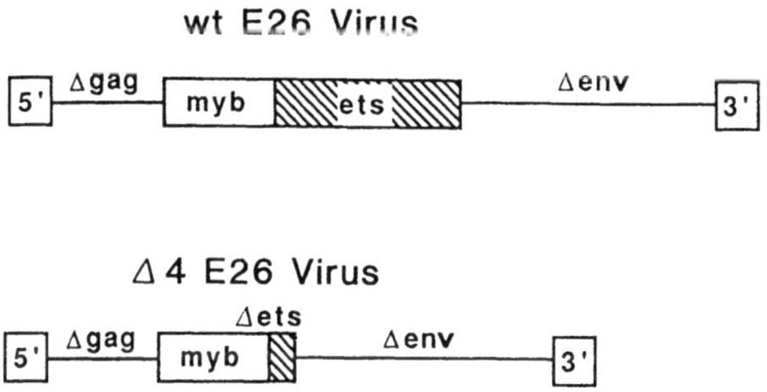

Fig. 1. Genome structure of wild-type E26 and of Δ4E26 viruses. The size of the *wt*E26 genome corresponds to 5.7 kb, that of Δ4E26 is shorter by 1266 internal base pairs of v-*ets* (Nunn et al. 1987)

[1]Differentiation Programme, European Molecular Biology Laboratory (EMBL), Meyerhofstrasse 1, 6900 Heidelberg, FRG

[2]Salk Institute, San Diego, California 92138, USA

38. Colloquium Mosbach 1987
Molecular Basis of Viral and Microbial Pathogenesis

Table 1. In vitro bone marrow transforming activity of Δ4E26 virus[a]

	Virus dilution 10°	10^{-1}
Δ4E26	180[b]	30[b]
*wt*E26	280	45

[a] 1×10^6 bone marrow cells prepared from a 20-day-old chick were infected with 0.5 ml of virus stock and seeded in Methocel as described earlier (Graf et al. 1981).

[b] Number of transformed colonies per dish.

Table 2. Leukemogenicity of Δ4E26 virus[a]

Virus	Incidence of animals with leukemia	Latency (days) Blast formation	Death
Δ4E26	0/10	-	-
*wt*E26	8/9	16.4	21.9

[a] 100 colony-forming units of virus were injected intravenously into 12 days of old embryos. The animals were observed over a period of 2 months.

This stock, when used to infect fresh bone marrow cells seeded in Methocel under myeloid conditions (Radke et al. 1982) had a titer of about 3.6×10^2 colony forming units (CFU) per ml and was thus comparable to that of *wt*E26 (Table 1). To determine the tumorigenicity of Δ4E26, a group of 25 twelve day old chick embryos were injected intravenously into the chorioallantoic membrane with 100 CFU of virus. As a control, another group of embryos was similarly injected with *wt*E26. After hatching, the animals were screened for the onset of leukemia by preparing blood smears twice weekly and staining the cells with benzidine and May Grünwald-Giemsa. Autopsies were performed on the animals that died. As shown in Table 2, none of the animals injected with Δ4E26 died with leukemia, while nearly all of those injected with E26 did. Buffy coat cells prepared from the peripheral blood of two leukemic animals 21 d after infection with *wt*E26 and stained with antibodies to histone 5 revealed, as expected, that the majority of the leukemic cells (82 and 92%, respectively) consisted of erythroblasts. No leukemic blasts could be found in the buffy coat from two animals infected with Δ4E26 for the same period of time. These results raised the question whether or not Δ4E26 virus had altogether lost its in vivo transforming capacity or whether low levels of leukemic cells were present in the blood which were not detected by visual inspection of the smears. We therefore incubated the buffy coat cells in liquid culture under myeloid conditions as described earlier (Radke et al. 1982). Rapidly proliferating myeloblasts appeared within 7 - 12 days in all the cultures,

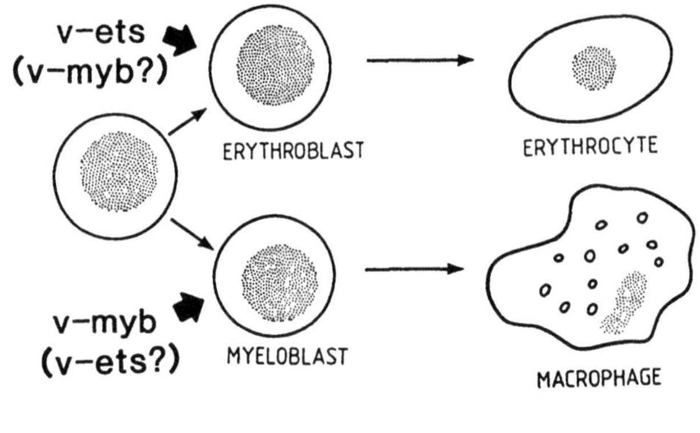

Fig. 2. Proposed target cells of the E26 virus oncogenes within the erythroid and myeloid lineages of chicken bone marrow

not only from the *wt*E26-infected but also from the Δ4E26-infected animals within 7 - 12 days, indicating that the mutant had retained the capacity to transform myeloid cells in vivo at low efficiencies. The results presented allow us to draw the following conclusions, schematically represented in the diagram of Fig. 2.

1. The v-*myb* gene is sufficient for myeloblast transformation. This result agrees with the observation that the v-*myb* only containing AMV virus transforms myeloblasts (Beug et al. 1979) and with our earlier studies on *ts* mutants of E26 that presumably map in v-*myb* (L. Frykberg, B. Vennström and T. Graf, unpublished observations). These mutants were found to be temperature-sensitive in their myeloid – but not in their erythroid – transforming activity (Beug et al. 1984). Whether the v-*ets* gene contributes to the transformed phenotype of v-*myb* transformed myeloblasts remains to be determined. It also remains unclear why AMV but not Δ4E26 is capable of inducing a lethal myeloblastosis in the animal. A possible explanation is provided by preliminary evidence indicating that AMV transformed cells are partly growth factor-independent while Δ4E26 and *wt*E26 cells are fully factor-dependent (T.G., unpublished results).
2. The v-*ets* gene is necessary for erythroid cell transformation since Δ4E26 is unable to induce an erythroleukemia. Whether v-*ets* alone is sufficient to do so cannot be decided at present. This will have to await the isolation and characterization of E26 mutants defective in the v-*myb* gene, a task which has turned out to be unexpectedly difficult, possibly because E26 proteins lacking *myb* are not stable.

In conclusion, E26 is a unique retrovirus that is capable of transforming hematopoietic cells of two different lineages, a property that correlates with its content of two distinct oncogenes. It thus differs from the two other known dual oncogene retrovirus isolates, the ES4 strain of avian erythroblastosis virus (AEV) containing v-*erbA* and v-*erbB* and the MH2 strain of myelocytomatosis virus containing v-*mil* and v-*myc*. These viruses selectively transform cells within the erythroid and myelomonocytic lineages, respectively (Kahn et al. 1986, Graf et al. 1986). It will now be interesting to determine how the v-*myb* and v-*ets* oncogenes cooperate in the transformation of erythroid and myeloid cells.

References

Beug H, Kirchbach v A, Döderlein G, Conscience JF, Graf T (1979) Chicken hematopoietic cells transformed by seven strains of defective avian leukemia viruses display three distinct phenotypes of differentiation. Cell 18:375-390

Beug H, Hayman MJ, Graf T (1982) Myeloblasts transformed by the avian acute leukemia virus E26 are hormone-dependent for growth and for the expression of a putative *myb*-containing protein, p135 E26. EMBO J 1:1069-1073

Beug H, Leutz A, Kahn P, Graf T (1984) *Ts* mutants of E26 leukemia virus allow transformed myeloblasts, but not erythroblasts or fibroblasts, to differentiate at the nonpermissive temperature. Cell 39:579-588

Graf T, Beug H (1978) Avian leukemia viruses: interaction with their target cells in vivo and in vitro. BBA Rev Cancer 516:269-299

Graf T, Stehelin D (1982) Avian leukemia viruses. Oncogenes and genome structure. BBA Rev Cancer 651:245-271

Graf T, Oker-Blom N, Todorov TG, Beug H (1979) Three new types of viral oncogenes of cellular origin specific for haematopoietic cell transformation. Nature (London) 281:452-455

Graf T, Kirchbach v A, Beug H (1981) Characterization of the hematopoietic target cells of AEV, MC29 and AMV avian leukemia viruses. Exptl Cell Res 131:331-343

Graf T, Kahn P, Leutz A, Vennström B, Beug H (1981) Lineage-specific oncogene cooperativity in avian hematopoietic cells. In: Cell cycle and oncogenes. Springer, Berlin Heidelberg New York, pp 131-134
Graf T, Weizsäcker F v, Grieser S, Coll J, Stehelin D, Patschinsky T, Bister K, Bechade C, Calothy G, Leutz A (1986) V-*mil* induces autocrine growth and enhanced tumorigenicity in v-*myc* transformed avian macrophages. Cell 100:357-364
Kahn P, Frykberg L, Brady C, Stanley IJ, Beug H, Vennström B, Graf T (1986) V-*erb*A cooperates with sarcoma oncogenes in leukemic cell transformation. Cell 100:349-356
Leutz A, Beug H, Graf T (1984) Purification and characterization of cMGF, a novel chicken myelomonocytic growth factor. EMBO J 3:3191-3197
Nunn MF, Duesberg PH, Hunter T (1987) The *ets* oncogene is required for transformation of erythroblasts by avian virus E26. (Submitted)
Radke K, Beug H, Kornfeld S, Graf T (1982) Transformation of both erythroid and myeloid cells by E26, an avian leukemia virus that contains the *myb* gene. Cell 31:643-653
Sotirov N (1981) Histone H5 in the immature blood cells of chickens with leukosis induced by avian leukosis virus strain E26. J Natl Cancer Inst 66:1143-1149

Mx Gene Control of Influenza Virus Susceptibility

O. Haller and P. Staeheli[1]

Summary

Alleles at the interferon-regulated *Mx* gene locus on chromosome 16 determine susceptibility of mice to influenza virus infections. Type I (alpha/beta) interferons transiently stimulate transcription of the *Mx* gene. Activation of the Mx^+ allele leads to the production and nuclear accumulation of the 72 kDa Mx protein and, concomitantly, to the establishment of a potent antiviral state against influenza viruses. Most laboratory mouse strains carrying deletions at the *Mx* locus (Mx^-) are unable to synthesize the Mx protein and are virus-susceptible. A full-length cDNA encoding Mx protein was cloned and constitutively expressed in susceptible Mx^- cells. Transfected cells that synthesized Mx protein resisted infection with influenza virus in the absence of any interferon treatment, demonstrating that the Mx protein is instrumental for influenza virus resistance. Mouse-adapted influenza virus strains grow rapidly in the lungs or other organs of infected Mx^- mice, thereby inducing large amounts of endogenous IFN. In similarly infected Mx^+ mice, only small necrotic lesions are found, which are self-limiting. By double immunofluorescence, we could demonstrate that these lesions contain viral antigen in the center and that the noninfected cells surrounding each lesion contain Mx protein in their nuclei. We conclude that local production of type I interferon in the affected organs and expression of Mx protein in the local tissue is essential for recovery from an otherwise lethal influenza virus infection.

Interferon-Mediated Genetic Resistance to Influenza Virus

Within the mouse species, instances are known in which a single gene determines great differences in susceptibility to defined groups of viruses (1). Resistance to the lethal action of mouse-adapted orthomyxoviruses could be attributed to the presence of the dominant Mx^+ allele at the *Mx* locus on chromosome 16 of the mouse (2,3,4). Mice bearing the Mx^+ allele were highly resistant to challenge with a variety of pneumotropic, neurotropic, and hepatropic influenza A viruses (3). This unusual resistance could not be explained by obvious mechanisms such as lack of virus receptors or a particularly efficient immune response. However, we found that virally induced type I (alpha/beta) interferon (IFN) was a necessary component of resistance since influenza virus resistance could be abolished by treatment of Mx^+ mice with neutralizing amounts of anti-interferon antibodies (5). It became clear that allelels at the *Mx* locus somehow influenced the protective effect of type I IFN against influenza viruses (6). Thus, purified mouse type I IFN was able to induce resistance to influenza virus in cultured cells of resistant Mx^+ animals but not in cells obtained from susceptible animals carrying Mx^- alleles. Mx^- cells were otherwise responsive to IFN: they were equally well protected by IFN against

[1]Institut für Immunologie und Virologie der Universität Zürich, Gloriastrasse 30, 8028 Zürich, Switzerland

38. Colloquium Mosbach 1987
Molecular Basis of Viral and Microbial Pathogenesis

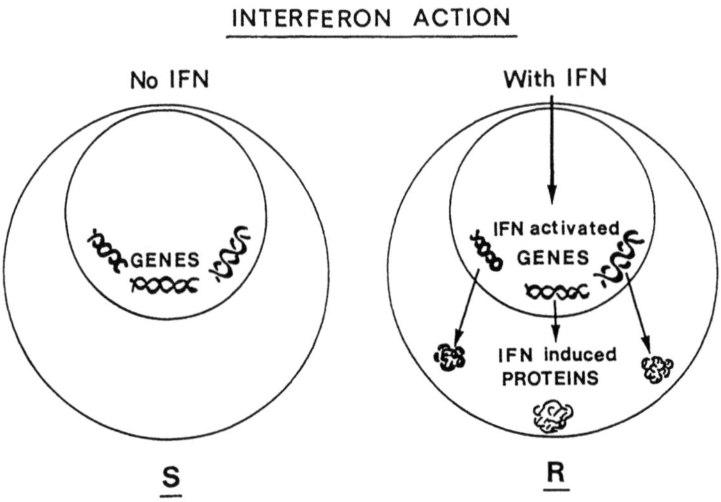

Fig. 1. Interferon action. IFN induces in cells the synthesis of several proteins which are thought to be mediators of the multiple effects of IFNs. Susceptible cells *(S)* become resistant *(R)* to most viruses

vesicular stomatitis virus or encephalomyocarditis virus as Mx^+ cells. We concluded that IFN was capable of inducing functionally different antiviral states in Mx^+ and Mx^- cells. Most likely, Mx^+ cells had a functionally intact *Mx* gene responsive to IFN, whereas Mx^- cells were defective.

Interferons Activate Cellular Genes

It is now well established that the various interferons all require active cellular RNA and protein synthesis for their action. So far, more than 15 IFN-induced proteins have been identified (7) and a current estimate of the number of IFN-activated genes is in the order of 15 - 20 for most cells (8). Clearly, synthesis of some of these gene products is required for the establishment of an antiviral state (Fig. 1). A first group of IFN-induced proteins represent enzymes of known functions. Best characterized are two enzymes originally purified from cell-free extracts of IFN-treated cells which need to be activated by double-stranded RNA: (2'-5') oligo A synthetase (9) and a protein kinase (10). A second group of IFN-induced polypeptides comprises proteins with no enzymatic activity but other known functions such as poly I:C (11) or guanylate binding proteins (12). Finally, polypeptides of a third class are presently defined by their position on gels following two-dimensional electrophoresis of cellular extracts and await further characterization. Antibodies have been produced to some of these proteins, which allow detection of their intracellular distribution.

Mx Protein Expression Correlates with Influenza Virus Resistance

In collaboration with Michel Horisberger, we found an IFN-induced 72 kDa protein in a two-dimensional gel system which proved to be unique to Mx^+ cells (13). This protein was not detectable in control Mx^+ cells that were kept in the absence of IFN, but was strongly induced in the same cells upon IFN stimulation. More importantly, this

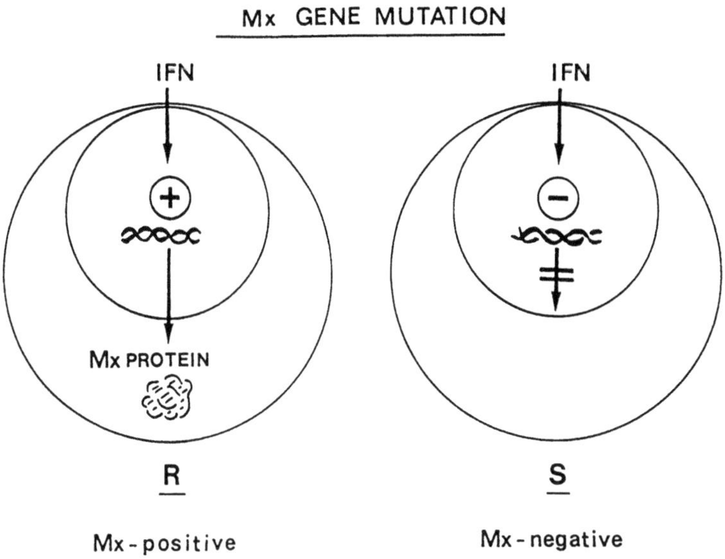

Fig. 2. Lack of proper IFN responsiveness due to host cell mutation. Cells with mutations at the IFN-activated *Mx* gene are unable to produce the Mx protein and remain susceptible to influenza virus despite exposure to IFN

protein was not inducible by IFN in Mx^- cells from influenza virus-susceptible mouse strains. We concluded that the newly detected protein was encoded by the Mx^+ allele and called it Mx protein. It has since become clear that Mx^- cells carry mutated versions of the Mx^+ allele which do not permit Mx protein synthesis (14, Peter Staeheli, unpublished). This situation is illustrated in Fig. 2.

Much of the present work relies on *Mx*-congenic animals. These were produced by crossing the Mx^+ resistance allele into current Mx^- laboratory mouse strains such as BALB/c and C57BL/6J (15). Each backcross generation was challenged with a suitable dose of influenza virus, which killed off all Mx^- animals but allowed Mx^+ carriers to survive. After 20 backcross generations, congenic resistant animals can be assumed to differ from their susceptible counterparts by only a small fraction of their genome surrounding the *Mx* locus on the distal part of chromosome 16. In all these crosses, inducibility of the Mx protein and the resistance phenotype co-segregated as a single trait.

Cloning and Expression of Mx cDNA: Recombinant Mx Protein is Inhibitory to Influenza Virus

We then immunized Mx^- mice with whole cell extracts of IFN-treated Mx^+ congenic cells and obtained polyclonal and monoclonal antibodies directed exclusively at the Mx protein (16). Immunofluorescence studies showed that the Mx protein accumulated in the nuclei of IFN-treated Mx^+ cells (17). IFN-treated Mx^- cells showed no specific staining. Untreated control cells of either genotype were negative (Fig. 3). Accordingly, Mx mRNA was expected to be present only in extracts from IFN-induced Mx^+ cells but not in IFN-induced extracts from Mx^- cells nor in extracts from non-induced cells. This fact, and the availability of convenient antibodies, facilitated the Mx cDNA cloning done in collaboration with Werner Boll and Charles Weissmann of the University of Zürich. A full-

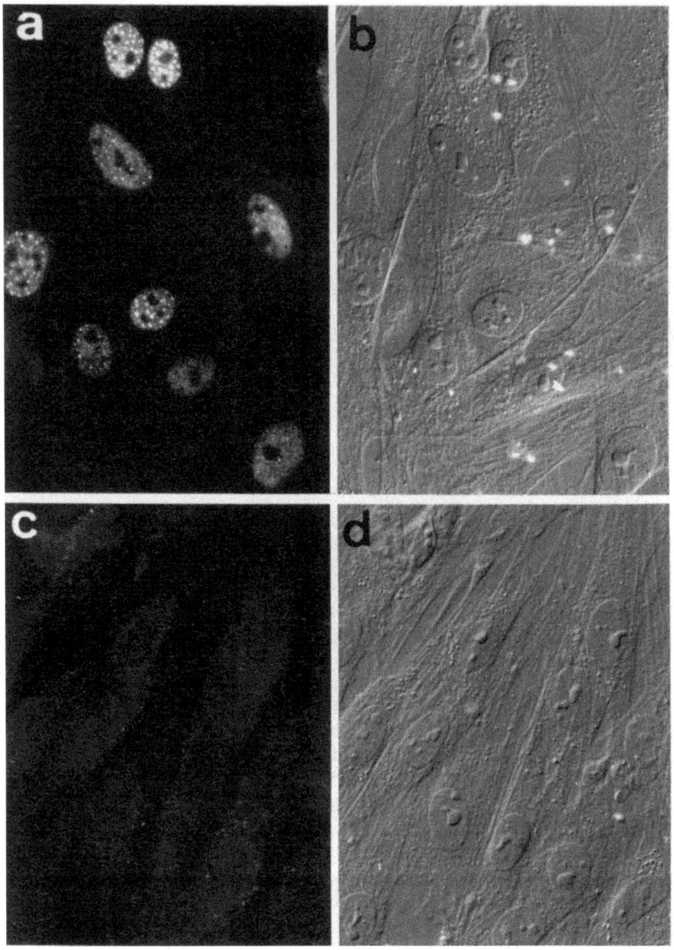

Fig. 3. Immunofluorescence staining of the nuclear *Mx* protein. Embryo cells from congenic Mx^+ (a, b) or Mx^- (c, d) mice were treated for 18 h with 300 reference units/ml of type I IFN and were then stained with a monoclonal anti-Mx antibody as described (17). Identical areas of each monolayer are shown by UV incident light fluorescence microscopy *(left)* or by Nomarski interference contrast optics *(right)*

length cDNA clone with an insert of 3.3 kb was obtained. Sequence analysis revealed a long open reading frame encoding a putative protein with 631 amino acids having no significant homology to other sequenced proteins. Constitutive expression of this cDNA in transfected Mx^- 3T3 cells resulted in nuclear accumulation of Mx protein and selective resistance of such cells against influenza virus even in the absence of IFN treatment. This indicated that the Mx protein alone, without cooperation from other IFN-induced proteins, was inhibitory to influenza virus (14).

Role of Mx Protein for Virus Resistance in Vivo

Extensive studies of the disease caused by influenza viruses in resistant and susceptible mice suggested early on that resistance was probably not mediated by specialized cells of the host defence system but, more likely, was a property of each target cell (18,19). A major difficulty in accepting IFN as a mediator of resistance was the fact that large amounts of IFN was usually found in infected tissues of susceptible Mx^- mice (5,20). Furthermore, IFN was believed to act against all

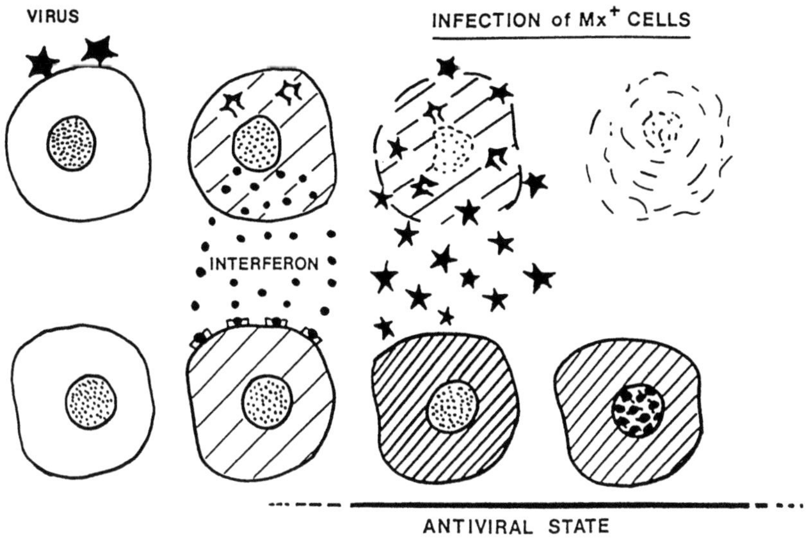

Fig. 4. *Mx* effect in vivo. IFN produced locally during infection activates the Mx^{+} allele and inhibits further influenza virus spread by establishing a potent antiviral state in cells of the surrounding tissue (see text)

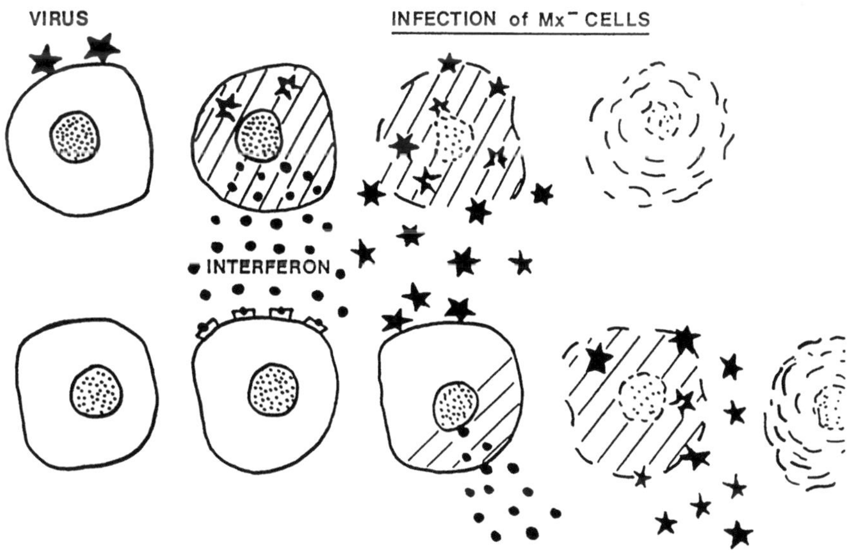

Fig. 5. Failure of IFN action against influenza virus in Mx^{-} tissues. In the absence of Mx protein induction cells remain virus susceptible. More cells are infected and more IFN is produced without much protective effect (see text)

viruses more or less indiscriminately. However, if we accept that the antiviral spectrum of IFN depends on proper activation of certain host cell genes and that susceptible mice carry nonfunctional mutations at the Mx locus, the following picture emerges. During the course of an influenza virus infection type I IFN is produced and secreted by cells exposed to the virus in the infected organs. This process occurs in *Mx*$^{+}$ as well as in *Mx*$^{-}$ animals. The local IFN then binds to appropriate receptors on the surface of neighboring cells and leads to the activation of a set of IFN-dependent genes. In the case of resistant animals, the *Mx*$^{+}$ allele is activated and Mx protein is produced which eventually accumulates in the cell nucleus. All cells expressing Mx protein are then protected from subsequent infection by the invading influenza virus. Consequently, virus spread comes to a halt and further cell damage is prevented (Fig. 4). In the case of *Mx*$^{-}$ cells, IFN acts in the same way, except that it is unable, for genetic reasons, to induce the synthesis of the Mx protein. Hence, no effective antiviral resistance is established in such cells and virus spread continues unhindered (Fig. 5). That this scenario is real can be shown in the following way. A hepatotropic influenza virus strain grows in hepatocytes and causes fulminant generalized liver necrosis and death in *Mx*$^{-}$ mice (21). In *Mx*$^{+}$ mice it produces only small necrotic foci which are self-limiting (16). Using double immunofluorescence techniques, we could recently demonstrate that these lesions contain viral antigen in the center and that the noninfected cells surrounding each lesion contain Mx protein in their nuclei (P. Juelke and O. Haller, in preparation). Similar findings were made with lungs or brains of mice infected with pneumotropic or neurotropic influenza virus strains indicating that the Mx resistance mechanism operates in different organs against different influenza viruses in much the same way.

Acknowledgments. We thank Heinz Arnheiter, Jean Lindenmann, and Charles Weissmann for stimulating discussions. The present work was supported by the Swiss National Science Foundation, Grants No. 3.507-083 and 3.507-086.

References

1. Brinton MA, Blank KJ, Nathanson N (1984) Host genes that influence susceptibility to viral diseases. In: Notkins AL, Oldstone MBA (eds) Concepts in viral pathogenesis. Springer, Berlin Heidelberg New York, pp 71-78
2. Lindenmann J (1964) Inheritance of resistance to influenza virus in mice. Proc Soc Exp Biol Med 116:506-509
3. Haller O (1981) Inborn resistance of mice to orthomyxoviruses. Curr Top Microbiol Immunol 92:25-52
4. Stäheli P, Pravtcheva D, Lundin L-G, Acklin M, Ruddle F, Lindenmann J, Haller O (1986) Interferon-regulated influenza virus resistance gene *Mx* is localized on mouse chromosome 16. J Virol 58:967-969
5. Haller O, Arnheiter H, Gresser I, Lindenmann J (1979) Genetically determined, interferon-dependent resistance to influenza virus in mice. J Exp Med 149:601-612
6. Haller O, Arnheiter H, Lindenmann J, Gresser I (1980) Host gene influences sensitivity to interferon action selectively for influenza virus. Nature (London) 283:660-662
7. Weil J, Epstein ChJ, Epstein LB, Sedmak JJ, Sabran JL, Grossberg SE (1983) A unique set of polypeptides is induced by gamma interferon in addition to those induced in common with alpha and beta interferons. Nature (London) 301:437-439
8. Revel M, Chebath J (1986) Interferon-activated genes. Trends Biochem Sci 11:166-170
9. St Laurent G, Yoshie O, Floyd-Smith G, Samanta H, Sehgal PB, Lengyel P (1983) Interferon action: Two (2'-5') $(A)_n$ synthetases specified by distinct mRNAs in Ehrlich ascites tumor cells treated with interferon. Cell 33:95-102

10. Sen GC, Taira H, Lengyel P (1978) Characteristics of a double-stranded RNA-activated protein kinase system partially purified from interferon-treated Ehrlich ascites tumor cells. J Biol Chem 253:5915-5921
11. Chebath J, Merlin G, Metz R, Benech Ph, Revel M (1983) Interferon-induced 56,000 M_r protein and its mRNA in human cells: Molecular cloning and partial sequence of the cDNA. Nucleic Acids Res 11:1213-1227
12. Prochazka M, Stäheli P, Holmes RS, Haller O (1985) Interferon-induced guanylate-binding proteins: Mapping of the murine *Gbp-1* locus to chromosome 3. Virology 145:273-279
13. Horisberger MA, Stäheli P, Haller O (1983) Interferon induces a unique protein in mouse cells bearing a gene for resistance to influenza virus. Proc Natl Acad Sci USA 80:1910-1914
14. Stäheli P, Haller O, Boll W, Lindenmann J, Weissmann C (1986) Mx protein: Constitutive expression in 3T3 cells transformed with cloned Mx cDNA confers selective resistance to influenza virus. Cell 44:147-158
15. Haller O, Prochazka M, Staeheli P, Lindenmann J (1984) *Mx*$^+$-congenic mouse strains. Mouse News Lett 71:53-54
16. Stäheli P, Dreiding Ph, Haller O, Lindenmann J (1985) Polyclonal and monoclonal antibodies to the interferon-inducible protein Mx of influenza virus resistant mice. J Biol Chem 260:1821-1825
17. Dreiding Ph, Stäheli P, Haller O (1985) Interferon-induced protein Mx accumulates in nuclei of mouse cells expressing resistance to influenza viruses. Virology 140:191-196
18. Haller O, Arnheiter H, Lindenmann J (1976) Genetically determined resistance to infection by hepatotropic influenza A virus in mice: Effect of immunosuppression. Infect Immun 13:844-854
19. Haller O, Arnheiter H, Lindenmann J (1979) Natural, genetically determined resistance toward influenza virus in hemopoietic mouse chimeras: role of mononuclear phagocytes. J Exp Med 150:117-126
20. Fiske RA, Klein PA (1975) Effect of immunosuppression on the genetic resistance of A2G mice to neurovirulent influenza virus. Infect Immun 11:576-581
21. Haller O (1975) A mouse hepatrophic variant of influenza virus. Arch Virol 49: 99-116

Shiga and Shiga-Like Toxins: A Family of Related Cytotoxins

A. D. O'Brien[1], M. P. Jackson[1], L. R. M. Marques[1], N. A. Strockbine[1,3], D. L. Weinstein[1], M. A. Moore[1], R. K. Holmes[1], J. W. Newland[2], and R. J. Neill[2]

Introduction

Shiga toxin is a potent cytotoxin produced by *Shigella dysenteriae* 1 that is also enterotoxic for ligated rabbit ileum and paralytic and lethal for mice and rabbits (1). Shiga toxin was first described in 1903 by Conradi (2), and it is the prototype toxin for a family of related toxins called Shiga-like toxins (SLT's) or verotoxins. The members of the SLT family include cytotoxins produced by other *Shigella* species (3,4), *Escherichia coli* (5,6) two species of *Vibrio* (7), as well as some strains of *Salmonella typhimurium* (5,8), and *Campylobacter jejuni* (M. Moore, M. Blaser, G. Perez-Perez, and A. O'Brien, in preparation). The levels of cytotoxic activity produced by these organisms as measured by the number of 50% cytotoxic doses (CD_{50}) per ml cell lysate (9) vary from low ($< 6 \times 10^2$) to moderate (1×10^3 to 1×10^4) to high (1×10^5) to 1×10^8). Production of moderate or high levels of SLT's by *E. coli* is commonly associated with strains of *E. coli* that cause hemorrhagic colitis or the hemolytic uremic syndrome (9,10). The significance (if any) of low-level SLT production remains to be determined. The purpose of this report is to summarize our current understanding of the epidemiology, potential role in pathogenesis, structure, function and molecular genetics of the various members of the SLT family. Particular emphasis will be placed on Shiga toxin and the SLT's produced at moderate to high levels by *E. coli*.

Shiga Toxin and Shiga-Like Toxins of Shigella

All wild-type strains of *S. dysenteriae* 1 (Shiga's bacillus) tested to date in our laboratory and by Prado et al. (11) produce high levels of Shiga toxin. This high-level toxin synthesis is unique among the shigellae; no other serotypes of *S. dysenteriae* (i.e., *S. dysenteriae*) serotypes 2-15) or other species of *Shigella* (i.e., *S. flexneri*, *S. boydii*, and *S. sonnei*) produce high levels of Shiga toxin (3,4,11). Among the shigellae, Shiga's bacillus most often causes the severest form of shigellosis, bacillary dysentery (12). *S. dysenteriae* 1 is also the serotype of *Shigella* most often associated with post-infection development of the hemolytic uremic syndrome (HUS), a life-threatening triad of hemolytic anemia, thrombocytopenia, and acute renal failure (13,14). It has been proposed that Shiga toxin contributes to the pathogenesis of bacillary dysentery by its direct cytotoxic action on colonic epithelial cells (15) and to the pathogenesis of HUS by its dissemination

[1]Department of Microbiology, Uniformed Services University of the Health Sciences, Bethesda, MD 20814-4799, USA

[2]Walter Reed Army Institute of Research, Washington, DC 20307-5100, USA

[3]Division of Bacterial Diseases, Centers for Disease Control, 1600 Clifton Road, Atlanta, GA 30333, USA

38. Colloquium Mosbach 1987
Molecular Basis of Viral and Microbial Pathogenesis

via the blood stream and subsequent cytotoxic action on vascular endothelial cells (10,15). In support of these hypotheses are the recent demonstrations that purified Shiga toxin is cytotoxic for human colonic epithelial cells maintained in vitro (16) and for human vascular endothelial cells isolated from umbilical cords (17). Shiga toxin may also stimulate net fluid secretion in the gut, and consequently, promote diarrhea by selectively destroying mature absorptive epitheal cells (18).

Like many bacterial toxins, Shiga toxin has an A:B subunit structure where the A subunit is responsible for enzymatic activity and the B subunit for binding to the eukaryotic cell receptor (reviewed in (19)). Shiga toxin is comprised of one A subunit (ca. 32,000 Da) and five copies (20) of a B subunit (ca. 7,700 Da). Shiga toxin binds to sensitive eukaryotic cells by a galactose $\alpha 1 \rightarrow 4$ galactose (galabiose)-containing glycolipid receptor (21,22), and enters the cell via receptor-mediated endocytosis (23). The enzymatically active protein of the A subunit is an A_1 fragment (ca. 27,000 Da) derived from the A subunit by proteolytic cleavage and reduction of sulfhydryl bonds (24). Whether the A_1 fragment is generated within the clathrin-coated pits or in the cytoplasm of the cell remains to be determined. The A_1 fragment inhibits protein synthesis in eukaryotic cells by catalytically inactivating 60S ribosomal subunits (25). The results of recent studies by Brown et al. (26) and Obrig et al. (27) indicate that the A_1 fragment prevents peptide chain elongation by altering elongation factor 1 (EF-1)-dependent aminoacyl t-RNA binding to the ribosomal subunit.

Studies on the genetics of Shiga toxin synthesis by *S. dysenteriae* 1 were only recently initiated in our laboratory because cloning of the intact structural genes for Shiga toxin was forbidden under the guidelines of the Recombinant DNA Advisory Committee of the National Institutes of Health until May of 1986 (28). The Shiga toxin structural genes have now been cloned under BL3 +EK1 conditions and nucleotide sequence data obtained (29). From the results of these experiments, we have concluded that Shiga toxin is chromosomally encoded and that only one copy of the structural genes for high-level toxin production exist on the *S. dysenteriae* 1 chromosome. We have also found that the nucleotide sequence of the A subunit of Shiga toxin is essentially identical to the nucelotide sequence that was recently reproted (30) for the A subunit of the phage-encoded SLT type I (SLT-I) of enterohemorrhagic *E. coli* 0157:H7. In addition, since the composition and sequence of the 69 amino acids of the Shiga toxin B sunit reported by Seidah et al. (31) are identical to the deduced amino acid sequence of the B subunit of SLT-I (30), it appears that Shiga toxin and SLT are essentially the same toxin.

Some strains of *S. flexneri* and *S. sonnei* produce low levels of a cytotoxin that can be neutralized by anti-Shiga toxin (3,4,11). The molecular basis for the variability in level of cytotoxin production among the shigellae has not been determined. However, it has been established that the differences in levels of cytotoxin produced by *S. flexneri* strain M4243 and several strains of *S. dysenteriae* 1 reflect variations in toxin yields rather than variations in cytotoxicity per unit of toxin antigen (32). Furthermore, results from Southern hybridization analyses (29)suggest that the structural genes for Shiga toxin and the genes for SLT production by *S. flexneri* are distinct but share some limited homology.

E. coli Shiga-Like Toxins

E. coli can produce: (a) high levels of SLT-I; (b) low levels of SLT-I; (c) moderate levels of SLT-II; (d) high levels of SLT-I and moderate levels of SLT-II; and (e) moderate levels of a variant of SLT-II. Elevated (i.e., moderate or high) levels of SLT-I, SLT-II, or SLT-I + II are produced mainly by strains that belong to serotypes recently classified (33) as enterohemorrhagic *E. coli* (EHEC), i.e., 0157:H7, 026:H11, and 0111:H8 (9,34).

EHEC 0157:H7 strain 933 which is the strain incriminated as the source of a food-borne outbreak of hemorrhagic colitis in 1982 (35) synthesizes both SLT-I and SLT-II (34). These SLT's of strain 933 have served as prototype toxins for our studies. SLT-I and SLT-II are encoded by distinct lysogenic phages (36) designated 933J and 933W, respectively. SLT-I is the dominant cell-associated cytotoxin in strain 933, whereas SLT-II predominates in the culture supernatant (34). The relative contribution of SLT-I and SLT-II to the pathogenesis of hemorrhagic colitis or the hemolytic uremic syndrome when a strain makes both toxins remains to be determined.

SLT-I (High Levels). SLT-I has been purified to homogeneity from EHEC strains of serotypes 026:H11 and 0157:H7 (37,38). The biological properties and subunit structures of the purified toxins are indistinguishable from purified Shiga toxin. The cytotoxic activity of SLT-I is neutralized by antiserum to Shiga toxin. Moreover, the eukaryotic receptor for SLT-I appears to be the same as the Shiga toxin receptor (39). These observations are not surprising in the light of the striking degree of nucleotide sequence homology between the structural genes of SLT-I (30) cloned from coliphage 933J (40) and Shiga toxin (see above).

The organization and expression of the SLT-I operon was inferred from analysis of the nucleotide sequence of SLT-I cloned from coliphage 933J (30). The structural genes for the A and B subunits of SLT-I (designated *slt*-I A and *slt*-I B) are oriented on a single transcriptional unit with *slt*-I A upstream of *slt*-I B. A 12-bp space separates the in-phase open reading frames of the genes. Putative ribosome binding sites are 5' to both the *slt*-I A and *slt*-I B genes. The ratio of one A subunit to 5 B subunits in the holotoxin may reflect more efficient translation of the B subunit. Translation of the SLT-I nucleotide sequence indicates that both the A and the B subunits are synthesized with signal peptides. The molecular weights of the mature A and B subunits as calculated from translation of the nucleotide sequences are 32,211 and 7690, resepctively.

SLT-I (Low Levels). Many strains of *E. coli* produce low levels of cell-associated cytotoxins (6,9). A significant proportion (40%) of these weakly reactive cytotoxins can be neutralized by anti-Shiga toxin (9). Included among these low-level SLT-I-producing *E. coli* are many normal gut flora strains, *E. coli* K-12 substrains, and some enteropathogenic *E. coli*, enteroinvasive *E. coli*, and enterotoxigenic *E. coli* (9). Indeed, we have speculated (9) that all *E. coli* strains may produce low levels of cytotoxin(s) but that our assay system may be too insensitive to consistently detect the toxin(s).

SLT-II (Moderate Levels). Purification of SLT-II to homogeneity has not as yet been reported. Crude preparations of SLT-II from an *E. coli* K-12 substrain lysogenized with the toxin-converting phage 933W exhibited the same biological properties as Shiga toxin and SLT-I: cytotoxicity for HeLa and Vero cells but not Chinese Hamster Ovary cells or Y-1 adrenal cells, enterotoxicity for ligated rabbit ileal segments and lethality for mice (34). However, antiserum to Shiga toxin does not neutralize the cytotoxic activity of SLT-II.

We recently deduced the molecular weights of what are presumed to be the A and B subunits of SLT-II from nucleotide sequence analysis of the structural genes cloned from coliphage 933W (41,42). The molecular weights of the mature SLT-II A and B polypeptides are predicted to be 33,135 and 7817. The operon structure of SLT-II is similar to that of SLT-I, except for the space between *slt*-A and *slt*-B. The spacing between the open reading frames of *slt*-II and *slt*-II B is 14 base pairs so that each subunit gene is translated in a different reading frame. This is in contrast to the in-frame *slt*-I A and *slt*-B structural genes which are separated by 12 base pairs (equivalent to 4 condons). There is extensive amino acid sequence homology between the A subunits and B subunits of SLT-I and SLT-II. However, the extent of sequence homology between SLT-I and SLT-II drops at the carboxy termini for both the A and B subunits. Overall, the A subunits of SLT-I and SLT-II share 56% nucleotide sequence homology and 49% amino acid sequence homology. The nucleotide and amino acid sequence homologies for the B subunits of SLT-I and SLT-II are 53% and 26%, respectively. Nonetheless, the secondary structure of the SLT-I and SLT-II B subunits is probably quite similar based on the observations that the cysteine residues are located in identical positions for both polypeptides and that the hydrophobicity-hydrophilicity plots are nearly superimposable. Thus, in spite of the fact that SLT-I and SLT-II are antigenically distinct toxins (i.e., not cross-neutralizable), they are remarkably similar in terms of deduced polypeptide structures and biological activities.

SLT-II Variant (Moderate Levels). Edema disease (ED) of swine is an often fatal disease of weanling pigs caused by intestinal infection with certain strains of *E. coli* that classically belong to serogroups 0138, 0139, or 0141 (43). The pigs display neurological symptoms such as staggered gait and limb paralysis, and on necropsy, edema of the stomach wall and colonic mesentery (44). It has been postulated that the characteristic clinical and histopathologic features of ED may reflect the action on the host of a heat-labile cytotoxin called the edema disease principle (45,46). Because the neurological manifestations of pigs with ED are similar to those seen in animals inoculated parenterally with Shiga toxin, we recently assessed the relationship between the edema disease principle (EDP) and SLT's (47). Culture supernatants of 72 out of 81 ED strains were cytotoxic for Vero but not HeLa cells. In contrast, culture filtrates of *E. coli* strains that make the prototpye SLT-I or SLT-II are cytotoxic for Vero and HeLa cells. Although cytotoxicity of ED strains was not neutralized by anti-Shiga toxin, it was neutralized by antiserum to SLT-II. None of the 20 toxigenic ED strains examined harbored toxin-converting phage. We concluded that the EDP is a variant of SLT-II because it is neutralized by anti-SLT-II but is not cytotoxic for HeLa cells.

The SLT-II variant has not as yet been purified. However, the toxin genes have been cloned and subcloned from a representative ED strain (D. Weinstein, R. Holmes, and A. O'Brien, unpublished data). The restriction enzyme map for the subclone of the SLT-II variant is similar but not identical to that of SLT-II. Sequencing of the SLT-II variant genes is in progress.

Conclusions

The various properties of the members of the Shiga and Shiga-like toxin family are summarized in Fig. 1. Although it is clear that the genes for Shiga toxin and SLT-I are essentially identical, how the

Toxin	Culture supern. cytotoxic for HeLa	Culture supern. cytotoxic for Vero	Level of cell-associated toxin (CD_{50}/ml)	Neutralized by α ST	Neutralized by MAbSLT-I	Neutralized by α SLT-II	Location of structural genes	Disease association
Shiga	+	+	~ $10^5 - 10^8$	+	+	–	Chrom.	Bacillary dysentery
SLT-I (high)	+	+	~ $10^5 - 10^8$	+	+	–	Phage	Diarrhea, HC, HUS
SLT-I (low)	–	–	~ 10^2	+	+	–	Chrom. ?	?
SLT-II	+	+	~ $10^3 - 10^4$	–	–	+	Phage	Diarrhea, HC, HUS
SLT-II VARIANT	–	+	~ $10^3 - 10^4$	–	–	+	Chrom. ?	Edema disease

Fig. 1. Characteristics of the Shiga and Shiga-like toxin family

genes were exchanged between *Shigella* and *E. coli* is unknown. The structural genes for STL-II seem to have evolved by a series of point mutations from either SLT-I or Shiga toxin. The degree of genetic homology between SLT-II and the SLT-II variant remains to be determined. A comparison of the nucleotide sequence of all the Shiga and Shiga-like toxin structural genes as well as the sequences flanking these genes should facilitate an understanding of the evolution of the members of this toxin family.

References

1. O'Brien AD, LaVeck GD, Griffin DE, Thompson MR (1980) Characterization of *Shigella dysenteriae* 1 (Shiga) toxin purified by anti-Shiga toxin affinity chromatography. Infect Immun 30:170-179
2. Conradi H (1903) Über lösliche, durch aseptische Autolyse erhaltene Giftstoffe von Ruhr- und Typhusbazillen. Dtsch Med Wochenschr 29:26-28
3. Keusch GT, Jacewicz M (1977) The pathogenesis of *Shigella diarrhea*. VI. Toxin and antitoxin in *Shigella flexneri* and *Shigella sonnei* infections in humans. J. Infect Dis 135:552-556
4. O'Brien AD, Thompson MR, Gemski P, Doctor BP, Formal SB (1977) Biological properties of *Shigella flexneri* 2A toxin and its serological relationship to *Shigella dysenteriae* 1 toxin. Infect Immun 15:796-798
5. O'Brien AD, LaVeck GD, Thompson MR; Formal SB (1982) Production of *Shigella dysenteriae* type 1-like cytotoxin by *Escherichia coli*. J Infect Dis 146:763-769
6. Cleary TG, Mathewson JJ, Faris E, Pickering LK (1985) Shiga-like cytotoxin production by enteropathogenic *Escherichia coli* serogroups. Infect Immun 47:335-337
7. O'Brien AD, Chen ME, Holmes RK, Kaper J, Levine MM (1984) Environmental and human isolates of *Vibrio cholerae* and *Vibrio parahaemolyticus* produce a *Shigella dysenteriae* 1 (Shiga)-like cytotoxin. Lancet 1:77-78
8. Strockbine NA, Marques LRM, Holmes RK, O'Brien AD (1985) Characterization of monoclonal antibodies against Shiga-like toxin from *Escherichia coli*. Infect Immun 50:695-700
9. Marques LRM, Moore MA, Wells JG, Wachsmuth IK, O'Brien AD (1986) Production of Shiga-like toxin by *Escherichia coli*. J Infect Dis 154:338-341
10. Karmali MA, Petric M, Lim C, Fleming PC, Arbus GS, Lior H (1985) The association between idiopathic hemolytic uremic syndrome and infection by Verotoxin-producing *Escherichia coli*. J Infect Dis 151:775-782
11. Prado D, Cleary TG, Pickering LK, Ericsson CD, Bartlett AV III, DuPont HL, Johnson PC (1986). The relationship between cytotoxin production and clinical features in shigellosis. J Infect Dis 154:149-155
12. Keusch GT, Donohue-Rolfe A, Jacewicz M (1982) *Shigella* toxin(s). description and role in diarrhea and dysentery. Pharmacol Ther 15:403-438

13. Koster F, Levin J, Walker L, Tung, Gilman RH, Rahaman MM, Majid MA, Islam S, Williams RC Jr (1977) Hemolytic-uremic syndrome after shigellosis. Relation to endotoxemia and circulating immune complexes. N Engl J Med 298:927-933
14. Gasser C, Gautier E, Steck A, Siebenmann RE, Oechslin R (1955) Hämolytisch-urämisches Syndrom: bilaterale Nierenrinden-Nekrosen bei akuten erworbenen hämolytischen Anämien. Schweiz Med Wochenschr 85:905-909
15. Cantey JR (1985) Shiga toxin - an expanding role in the pathogenesis of infectious diseases. J Infect Dis 151:766-771
16. Moyer MP, Rothman SW, Brown JE (1985) Cytotoxicity of Shiga toxin for human colonic and ileal epithelial cells in culture. Abstr Annu Meet Am Soc Microbiol B-108:36
17. Obrig T, Del Vecchio PJ, Brown JE, Moran TP (1987) Direct cytotoxic activity of Shiga toxin on human vascular endothelial cells in culture. Abstr Annu Meet Am Soc Microbiol B-69:36
18. Keenan KP, Sharpnack DD, Collins H, Formal SB, O'Brien AD (1986) Morphologic evaluation of the effects of Shiga toxin and *E. coli* Shiga-like toxin on the rabbit intestine. Am J Pathol 125:69-80
19. O'Brien AD, Holmes RK (1987) Shiga and Shiga-like toxins. Microbiol Rev 51: 206-220
20. Donohue-Rolfe A, Keusch GT, Edson C, Thorley-Lawson D, Jacewicz M (1984) Pathogenesis of *Shigella* diarrhea. IX. Simplified high yield purification of *Shigella* toxin and characterization of subunit composition and function by the use of subunit-specific monoclonal and polyclonal antibodies. J Exp Med 160:1767-1781
21. Brown JE, Karlsson KA, Lindberg A, Stromberg N, Thurin J (1983) Identification of the receptor glycolipid for the toxin of *Shigella dysenteriae*. In: Chester A, Heinegard D, Lundblad A, Svensson S (eds) Glycoconjugates. Proc 7th Int Symp Glycoconjugates. Lund-Ronneby, Sweden, pp 678-679
22. Lindberg AA, Brown JE, Strömberg N, Westling-Ryd M, Schultz JE, Karlson KA (1987) Identification of the carbohydrate receptor for Shiga toxin produced by *Shigella dysenteriae* type 1. J Biol Chem (in press)
23. Jacewicz M, Keusch GT (1983) Pathogenesis of *Shigella* diarrhea. VII. Evidence for a translocation step in the cytotoxic action of Shiga toxin. J Infect Dis 148:844-854
24. Olsnes S, Reisbig R, Eiklid K (1981) Subunit structure of *Shigella* cytotoxin. J Biol Chem 256:8732-8738
25. Reisbig R, Olsnes S, Eiklid K (1981) The cytotoxic activity of *Shigella* toxin. Evidence for catalytic inactivation of the 60S ribosomal subunit. J Biol Chem 256:8739-8744
26. Brown JE, Obrig TG, Ussery MA, Moran TP (1986) Shiga toxin from *Shigella dysenteriae* 1 inhibits protein synthesis in reticulocyte lysates by inactivation of aminoacyl-tRNA binding. Microb Pathogenesis 1:325-334
27. Obrig TG, Moran TP, Brown JE (1987) The mode of action of Shiga toxin on peptide elongation of eucaryotic protein synthesis. J Biol Chem (in press)
28. Federal Register (1986) 51:16972
29. Strockbine NA, Jackson MP, Sung LM, Holmes RK, O'Brien AD (1987) Cloning and sequencing of the genes for Shiga toxon from *Shigella dysenteriae* type 1. J Bacteriol (in press)
30. Jackson MP, Newland JW, Holmes RK, O'Brien AD (1987) Nucleotide sequence analysis of the structural genes for Shiga-like toxin I encoded by bacteriophage 933J from *Escherichia coli*. Microb Pathogenesis 2:147-153
31. Seidah NC, Donohue-Rolfe A, Lazure C, Auclair F, Keusch GT, Chrêtein M (1986) Complete amino acid sequence of *Shigella* toxin B-chain. A novel polypeptide containing 69 amino acids and one disulfide bridge. J Biol Chem 261:13928-13931
32. O'Brien AD, LaVeck GD (1982) Immunochemical and cytotoxic activities of *Shigella dysenteriae* 1 (Shiga) and Shiga-like toxins. Infect Immun 35:1151-1154
33. Levine MM (1987) *Escherichia coli* that cause diarrhea: enterotoxigenic, enteropathogenic, enteroinvasive, enterohemorrhagic, and enteroadherent. J Infect Dis 155:377-389
34. Strockbine NA, Marques LRM, Newland JW, Smith HW, Homes RK, O'Brien AD (1986) Two toxin-converting phages from *Escherichia coli* O157:H7 strain 933 encode antigenically distinct toxins with similar biological activities. Infect Immun 53:135-140

35. Riley LW, Remis RS, Helgerson SD, McGee HB, Wells JG, Davis BR, Hebert RJ, Olcott ES, Johnson LM, Hargrett NT, Blake PA, Cohen ML (1983) Hemorrhagic colitis associated with a rare *Escherichia coli* serotype. N Engl J Med 308:681-685
36. O'Brien AD, Newland JW, Miller SF, Holmes RK, Smith HW, Formal SB (1984) Shiga-like toxin-converting phages from *Escherichia coli* strains that cause hemorrhagic colitis or infantile diarrhea. Science 226:694-696
37. O'Brien AD, LaVeck GD (1983) Purification and characterization of a *Shigella dysenteriae* 1-like toxin produced by *Escherichia coli*. Infect Immun 40:675-683
38. O'Brien AD, Lively TA, Chang TW, Gorbach SL (1983) Purification of *Shigella dysenteria* 1 (Shiga-like) toxin from *Escherichia coli* 0157:H7 strain associated with hemorrhagic colitis. Lancet 2:573
39. Lingwood CA, Law H, Richardson S, Petric M, Brunton JL, Karmali M (1987) Glycolipid binding of natural and cloned *Escherichia coli* produced Verotoxin in vitro. J Biol Chem (in press)
40. Newland JW, Strockbine NA, Miller SF, O'Brien AD, Holmes RK (1985) Cloning of Shiga-like toxin structural genes from a toxin converting phage of *Escherichia coli*. Science 230:179-181
41. Jackson MP, Neill RJ, O'Brien AD, Holmes RK, Newland JW (1987) Nucleotide sequence analysis and comparison of the structural genes for Shiga-like toxin I and Shiga-like toxin II encoded by bacteriophages from *Escherichia coli* 933. FEMS Lett 44:109-114
42. Newland JW, Strockbine NA, Neill RJ (1987) Cloning of genes for production of *Escherichia coli* Shiga-like toxin type II. Infect Immun 55:2675-2860
43. Sojka WJ, Lloyd MK, Sweeney EJ (1960) *Escherichia coli* serotypes associated with certain pig diseases. Res Vet Sci 1:17-27
44. Timoney JF (1950) Oedema disease of swine. Vet Rec 49:748-755
45. Nielsen NO, Clugston RE (1971) Comparison of *E. coli* endotoxin shock and acute experimental edema disease in young pigs. Ann Ny Acad Sci 176:176-189
46. Dobrescu L (1983) New biological effect of edema disease principle (*Escherichia coli*-neurotoxin) and its use as an in vitro assay for this toxin. Am J Vet Res 44:31-34
47. Marques LRM, Peiris JSM, Cryz SJ, O'Brien AD (1987) *Escherichia coli* strains isolated from pigs with edema disease produce a variant of Shiga-like toxin II. FEMS Lett 44:33-38

Molecular Mechanisms of Pathogenicity in *Shigella flexneri*

P. J. Sansonetti[1], B. Baudry[1], P. Clerc[1], A. T. Maurelli[1], X. Nassif[1], and A. Ryter[2]

Introduction

Although shigellosis is endemic throughout the world, it is of special concern in developing areas in which poor sanitation and low personal hygiene standards are responsible for a large number of cases. Children represent the major target of the disease, and malnutrition certainly accounts for the severity of the cases as well as a significant rate of mortality.

The primary step in the pathogenesis of bacillary dysentery is invasion of the human colonic mucosa (1). The invasive process encompasses complex features which include penetration into epithelial cells, intracellular multiplication, spreading to adjacent cells and to the conjunctive tissue of the intestinal villus (2). This sequence of events leads to a strong inflammatory reaction which causes ulcerations and abscesses (1) but may also prevent systemic dissemination of the pathogen. Therefore, the overall invasive process can be considered as the integration of two steps: one of invasion of individual epithelial cells and one of tissue invasion. The former requires the ability of the bacterium to induce its own phagocytosis by cells which are non professional phagocytes. It can be approximated in vitro by infecting monolayers of mammalian cells such as Henle intestinal or HeLa cells. The second step of tissue invasion requires survival of the bacteria within tissues, elicitation of the inflammatory reaction and tissue destruction. It can be studied in more definitive assays such as the Sereny test (3), and the rabbit ligated ileal loop (4).

Location of Genetic Sequences Necessary for Virulence in *Shigella Flexneri*

Plasmids of 140 MDa are consistently found in *S. flexneri* (5) as well as in other *Shigella* species and enteroinvasive *E. coli* (6,7,8). These plasmids are associated with the invasive phenotype, they are largely homologous and certainly derived from a common ancestor. pWR100, the virulence plasmid of *S. flexneri* serotype 5 has been further studied.

Homology between *E. coli* and *Shigella* is so high that they cannot be differentiated on the basis of polynucleotide hybridization. Therefore, conjugal mating between these genera allowed construction of hybrid strains by a stepwise transfer of previously identified virulence-associated genes from *S. flexneri* 2a Hfr to a rough, avirulent, *E. coli* K12 recipient (9). This process showed that the 140 MDa plasmid pWR100 was sufficient to confer the ability to invade HeLa cells. In addition, *Shigella* chromomosal genes are necessary for expression of a full virulence phenotype when hybrids are tested in more definitive assays. These

[1]Service des Entérobactéries, Unité INSERM 199
[2]Unité de Microscopie Electronique
Institut Pasteur, 28 rue du Docteur Roux, 75724 Paris Cedex 15, France

38. Colloquium Mosbach 1987
Molecular Basis of Viral and Microbial Pathogenesis

chromosomal segments include the *his*$^+$ region (45 min), which had already been linked to the expression of the somatic antigen (10). A region designated *kcp*+ which is linked to *pur*E (12 min) had already been shown to be necessary for a positive Sereny test (11). Finally, a poorly defined region limited by *mtl*+ (80 min) and *arg*+ (87 min) had been shown to be required for survival of strains in the intestinal mucosa (12). It must be included in *E. coli* hybrids which eventually invade the rabbit ileal loop with tissue destruction, elicit a positive Sereny test and cause fluid accumulation.

Phenotypic Expression of the Chromosomal and Extrachromosomal Virulent Loci

Role of Plasmid pWR100. pWR100 encodes functions that are essential for invasion of mammalian cells including entry into cells, intracellular multiplication, and killing of host cells. Thus, essential steps of the infectious process are under the control of extrachromosomal genes.

Recent evidence indicates that *S. flexneri* penetrates into cells via directed phagocytosis (P. Clerc, P.J. Sansonetti, submitted). This was suggested by the observation that F-actin, as well as myosin, accumulated underneath the cell membrane at the site of entry. Plasmid DNA sequences involved in entry of *S. flexneri* into HeLa cells have been cloned in the cosmid vector pJB8 (13). The cloned inserts which restored invasiveness shared a common sequence of 37 kb. Expression of one of these recombinant plasmids, pHS4108, has been studied in minicells. It encodes 12 polypeptides which are resolved by two-dimensional electrophoresis (B. Baudry et al., submitted). Among these, Western blot analysis, using the serum of a monkey convalescent from shigellosis, revealed that four polypeptides were recognized as in the parental virulent strains M90T (13). Hale et al. have shown that the same serum recognized four equivalent polypeptides in all *Shigella* species and in enteroinvasive *E. coli*. These four polypeptides are the only ones that are regularly recognized during natural infection in human beings (14). Molecular weights of these polypeptides vary between 80 and 35 kDa. Although their precise role in the infection process is as yet unclear, they may turn out to be essential components of a future *Shigella* vaccine.

Involvement of pWR100 in Intracellular Multiplication. Plasmid associated genes are also needed for efficient intracellular multiplication (15). pWR100 endows *E. coli* K12 with the same replication potential as the wild-type *S. flexneri* M90T. On the other hand, the cosmid recombinant molecule pHS4108 which codes for penetration into cells does not confer on recipient strains the same ability to multiply intracellularly. Upon transfer of this plasmid in BS176, a plasmidless avirulent derivative of M90T, transformants invaded HeLa cells but replicated poorly intracellularly. Intracellular growth was as poor as *Salmonella typhimurium* C5. Such experiments also demonstrated a lack of correlation between intracellular multiplication and production of Shiga toxin. A sequential electron microscopic study performed on HeLa cells respectively infected with the four strains already mentioned demonstrated that M90T induced early lysis of the phagocytic membrane after penetration within cells. By 30 min after centrifugation-induced penetration, all bacteria were lying freely within the cytoplasm of the cells. Similar observations could be made with *E. coli* K12 carrying pWR100. On the other hand, BS176 (pHS4108) and *S. typhimurium* C5 demonstrated late and unefficient ability to lyze the phagocytic membrane. These electron microscope observations established a clear correlation between plasmid-mediated intracellular multiplication and ability to perform early lysis of the membrane surrounding the phagocytic vacuole. Demonstration of a plasmid-mediated contact hemolytic activity produced by virulent shigellae provides a likely candidate for lysis of the phagosome.

Involvement of Plasmid pWR100 in Early Killing of Host Cells. Using the continuous macrophage cell-line J774 to study the intracellular fate of both an invasive and a noninvasive, plasmidless derivative of *S. flexneri* M90T, plasmid pWR100 appeared to mediate a very efficient and rapid killing activity on host cells. Expression of this activity required bacteria to be intracellular, since macrophages were protected by cytochalasin D. A battery of strains differing in their levels of Shiga-like toxin production demonstrated that inhibition of protein synthesis was not required for early killing. Damage to macrophages rather correlated with the ability of invasive bacteria to rapidly and efficiently lyze the membrane of the phagocytic vacuole. The role of the release of bacteria within the cytosol for subsequent expression of cytotoxic activity is likely. Mitochondria may be the target for this activity (16).

Phenotypic Expression of Chromosomal Loci. Most shigellae seem to encode the O-antigenic determinants at a chromosomal locus closely linked to the histidine genes at 45 min on the map. *S. flexneri* contains an additional locus next to the proline genes, at 6 min, encoding the type-specific antigen (10). Expression of a smooth LPS is not necessary for invasion of cells (2,17) but rather for intraluminal survival of bacteria and protection against bactericidal activity of the serum within tissues.

In two species, plasmids are involved in LPS biosynthesis. In *S. sonnei*, a 120 MDa plasmid carries both the genes for invasion of eukaryotic cells and for biosynthesis of LPS O-side chains (6,18). In *S. dysenteriae* 1, a small 6 MDa plasmid encodes functions necessary for O-antigen expression (19).

Another critical factor for bacterial virulence is the availability of iron, which is essential for bacterial growth. The level of free iron is severely restricted in tissues, due to the fact that most of ferric ions are chelated both in the intracellular and extracellular compartments. Extracellular iron is attached to high-affinity iron-binding proteins: lactoferrin or transferrin. Pathogenic bacteria multiply in such hostile conditions if they produce siderophores which may scavenge iron from binding proteins.

S. flexneri produces the hydroxamate type siderophore aerobactin (20). Recent studies with the *E. coli* K 12-*S. flexneri* hybrids showed that one of the chromosomal segments located between *mtl*+ and *arg*+ which allowed bacteria to multiply successfully within the mucosa of the rabbit ileal loop contained the genes necessary for synthesis of aerobactin as well as of the ferric-aerobactin outer membrane receptor (21). In a recent series of experiments (X. Nassif et al., submitted), we have demonstrated that an *iuc*::Tn*10* mutant of *S. flexneri* M90T which no longer produced hydroxamate was neither impaired in its capacity to grow within the lumen of rabbit ligated intestinal loops nor limited in its capacity to grow intracellularly and kill cells. It was thus concluded that aerobactin production acted as a virulence factor for invasive bacteria at the stage of multiplication within tissues when lying within the extracellular compartment of the intestinal villus.

Finally, when constructing *E. coli* K 12-*S. flexneri* hybrids, extension of the sequence to LDC-allowed production of fluid within the intestinal loop, thus suggesting that genes encoding or regulating Shiga-toxin production had been transferred (9). It is as yet unclear whether Shiga toxin production endows bacteria with a higher virulence potential in term of tissue destruction or just codes for enterotoxicity. Production of a *tox*-mutant is necessary to resolve this critical problem.

Conclusion

Genetic and molecular characterization of *Shigella flexneri* major virulence factors should help in understanding the complex pathogenic process of shigellosis. In addition, these information should be of primary importance for the design of vaccine strains.

References

1. La Brec EH, Schneider H, Magnani TJ, Formal SB (1964) Epithelial cell penetration as an essential step in the pathogenesis of bacillary dysentery. J Bacteriol 88: 1503-1518
2. Formal SB, Hale TL, Sansonetti PJ (1983) Invasive enteric pathogens. Rev Infect Dis 5:5702-5707
3. Sereny B (1957) Experimental keratoconjunctivitis shigellosa. Acta Microbiol Acad Sci Hung 4:367-376
4. Formal SB, Kundel D, Schneider H, Kunev N, Sprinz H (1961) Studies with *Vibrio cholerae* in the ligated loop of the rabbit intestine. Br J Exp Pathol 42:504-510
5. Sansonetti PJ, Kopecko DJ, Formal SB (1982) Involvement of a plasmid in the invasive ability of *Shigella flexneri*. Infect Immun 35:852-860
6. Sansonetti PJ, Kopecko DJ, Formal SB (1981) *Shigella sonnei* plasmids: evidence that a large plasmid is necessary for virulence. Infect Immun 34:75-83
7. Sansonetti PJ, d'Hauteville H, Ecobichon C, Pourcel C (1983) Molecular expression of virulence plasmids in *Shigella* and entero-invasive *Escherichia coli*. Ann Microbiol (Paris) 134 A:295-318
8. Sansonetti PJ, d'Hauteville H, Formal SB, Toucas M (1982) Plasmid mediated invasiveness of "*Shigella*-like" *Escherichia coli*. Ann Microbiol (Paris) 133:351-355
9. Sansonetti PJ, Hale TL, Dammin GI, Kapper C, Collins HH, Formal SB (1983) Alterations in the pathogenicity of *Escherichia coli* K12 after transfer of plasmid and chromosomal genes from *Shigella flexneri*. Infect Immun 39:1392-1402
10. Formal SB, Gemski P Jr, Baron LS, La Brec EH (1970) Genetic transfer of *Shigella flexneri* antigens to *Escherichia coli* K12. Infect Immun 1:279-287
11. Formal SB, Gemski P Jr, Baron LS, La Brec EH (1971) A chromosomal locus which controls the ability of *Shigella flexneri* to evoke keratoconjunctivitis. Infect Immun 3:73-79
12. Formal SB, La Brec EH, Kent TH, Falkow S (1965) Abortive intestinal infection with an *Escherichia coli-Shigella flexneri* hybrid strain. J Bacteriol 89:1374-1382
13. Maurelli AT, Baudry B, d'Hauteville H, Hale TL, Sansonetti PJ (1985) Cloning of plasmid DNA sequences involved in invasion of HeLa cells by *Shigella flexneri*. Infect Immun 49:164-171
14. Hale TL, Oaks EV, Formal SB (1985) Identification and antigenic characterization of virulence-associated, plasmid-coded proteins of *Shigella* spp. and enteroinvasive *Escherichia coli*. Infect Immun 50:620-629
15. Sansonetti PJ, Ryter A, Clerc P, Maurelli AT, Mounier J (1986) Multiplication of *Shigella flexneri* within HeLa cells: lysis of the phagocytic vacuole and plasmid-mediated contact hemolysis. Infect Immun 51:461-469
16. Clerc PL, Ryter A, Mounier J, Sansonetti PJ (1987) Plasmid-mediated early killing of eukaryotic cells by *Shigella flexneri* as studied by infection of J774 macrophages. Infect Immun 55:521-527
17. Okamura N, Nakaya R (1977) Rough mutant of *Shigella flexneri* 2a that penetrates culture cells but does not evoke keratoconjunctivitis in guinea-pigs. Infect Immun 17:4-8
18. Kopecko DJ, Washington O, Formal SB (1980) Genetic and physical evidence of plasmid control of *Shigella sonnei* form I cell surface antigen. Infect Immun 29:207-214
19. Watanabe H, Timmis KN (1984) A small plasmid in *Shigella dysenteriae* 1 specifies one or more functions essential for O-antigen production and bacterial virulence. Infect Immun 43:391-396

20. Payne SM, Niesel DW, Peixotto SS, Lawlor KM (1983) Expression of hydroxamate and phenolate siderophores by *Shigella flexneri*. J Bacteriol 155:949-955
21. Griffiths E, Stevenson P, Hale TL, Formal SB (1985) Synthesis of aerobactin and a 76,000 dalton iron-regulated outer membrane protein by *Escherichia coli* K12-*Shigella flexneri* hybrids. Infect Immun 49:67-71

Antigenic Variation of Proteins on the Surface of Pathogenic *Neisseriae*

T. F. Meyer, R. Haas, and A. Stern[1]

The gram-negative coccus *Neisseria gonorrhoeae* is a major cause of sexually transmitted disease in humans. This bacterium represents a model for several intriguing virulence mechanisms. Most striking is its tremendous capacity to vary its major surface antigens, assumed to be an effective means for evading the immune surveillance of the infected human host. Antigenic variation of the surface components may further facilitate efficient adaptation of the pathogen to environmental changes in the host during the course of an infection. Of the known surface components pilin, the subunit component of the pili, and opacity protein (Op or PII) of gonococci, as well as the Op-related class 5 proteins of meningococci, have received particular attention (1-3). These proteins seem to play a central role in colonization of the pathogens in the human mucosal tissue (4,5). The expression of pili and Op can be turned on and off (phase variation), and a single cell can give rise to cells that produce structurally and antigenically different proteins (antigenic variation). In addition, more than one variant protein can be produced at a time. The total number of variant proteins expressed appears to be strikingly large. To assess the molecular principles underlying this complex phenomenons, we have isolated and characterized several genes coding for pilin and Op (7,8).

The Pilin Genes of *Neisseria gonorrhoeae*

In our molecular studies we could demonstrate that the genome of *N. gonorrhoeae* harbors multiple gene loci for pilin. Most of these *pil* genes contain variant sequence information (9,10), but only one or two loci contain a complete structural gene (*pil*E) with a promoter sequence. The *pil*E genes are constitutively transcribed and translated (11), while all other *pil* genes are silent (12), lacking the promoter and the conserved amino terminal coding sequences.

From a total of about eight *pil* loci we have mapped five loci in a chromosomal segment of approximately 50 kb in strain MS11. This includes the expression loci *pil*E1 and *pil*E2, and the major silent locus *pil*S1 which is located upstream of the expression locus *pil*E1 (Fig. 1). All mapped loci are arranged in the same orientation with regard to the open reading frame. In addition, all of them possess a 3'-terminal *Cla*I/*Sma*I segment. Both expression loci harbor one complete structural gene and several upstream truncated silent gene copies in tandem arrangement. Similarly, some silent loci contain more than one gene copy tandemly arranged and, as in case of the large *pil*S1 locus, linked by a short repetitive sequence (Fig. 1). Most silent copies (SC) start with codon position 44, lacking those sequences which code for the conserved amino-terminal region of gonococcal pilin. Other SC's begin with codons equivalent to amino acid positions 31 or 92; the smallest copies found in both expression loci start only after codon (Fig. 1). On the

[1]Max-Planck-Institut für Biologie, Spemannstrasse 34, D-7400 Tübingen, FRG

38. Colloquium Mosbach 1987
Molecular Basis of Viral and Microbial Pathogenesis

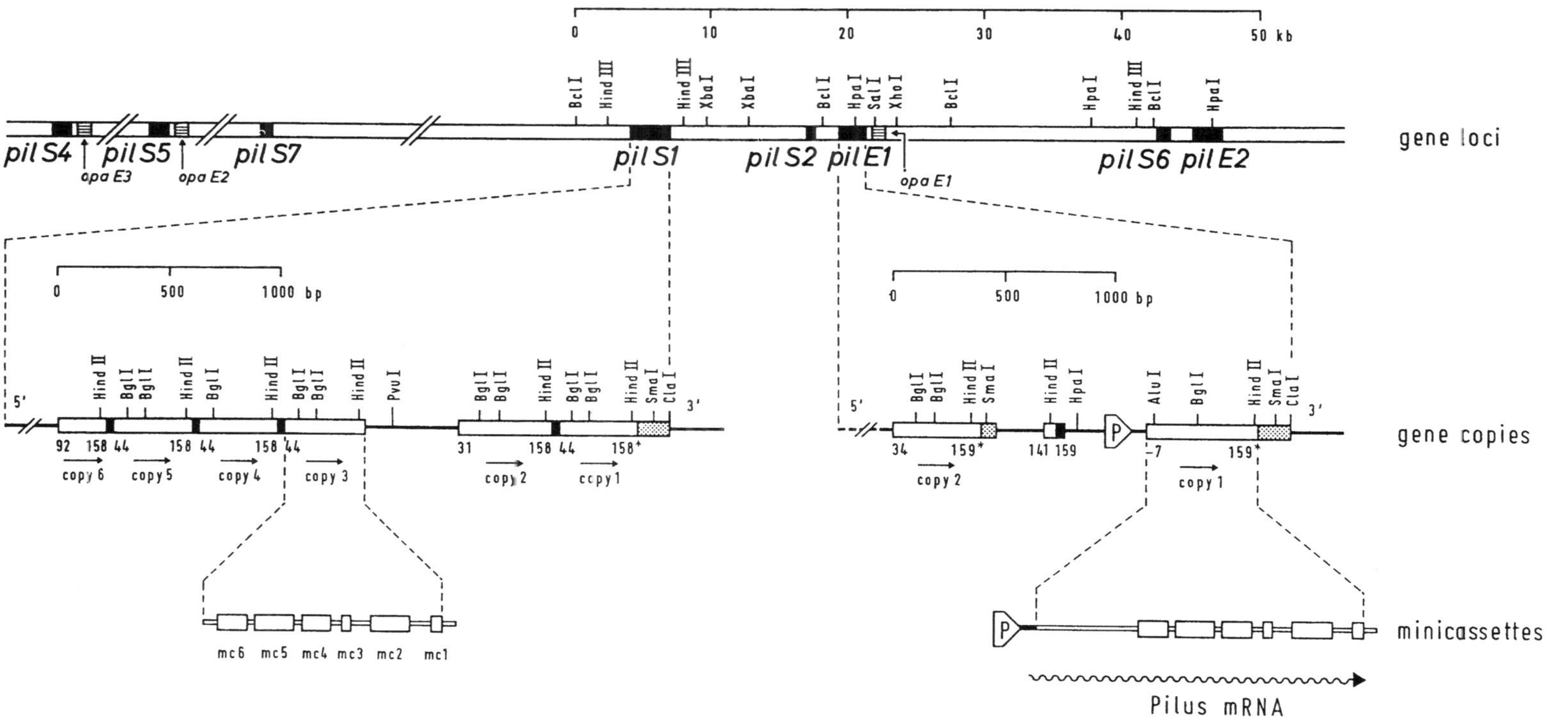

Fig. 1. Genetic map of pilus and opacity loci in *Neisseria gonorrhoeae* strain MS11. Besides the two expression loci (*pil*E1, *pil*E2) several silent loci (*pil*S) are located in a chromosomal segment of 50 kb. Some *pil* loci are closely linked with opacity loci (*opa*E). Expression loci contain one functional pilin gene copy (EC) and several 5'-truncated silent copies (SC), while silent loci are composed of one or several SC's. Silent and expressed pilin copies show the same characteristic minicassette structure. A pilin gene copy consists of six variable minicassettes (mc1-mc6) which are flanked by conservative sequences

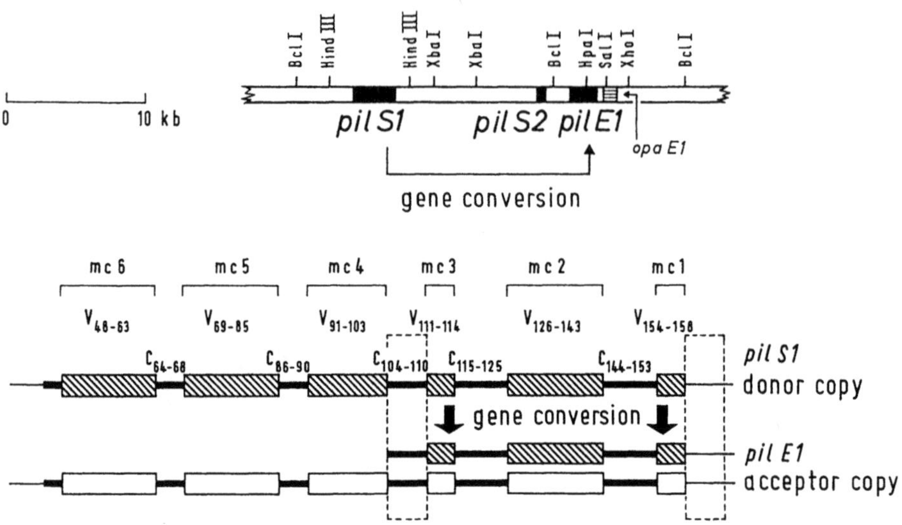

Fig. 2. Gene conversion of the expression gene in *pil*E1. In this example the *pil*E1 locus was converted by three minicassettes from the silent locus *pil*S1 located about 10 kb upstream *(top panel)*. Conserved regions *(C, solid lines* in the *bottom panel)* are assumed to support the recombination between the donor copy and the acceptor copy. The model is based on the sequence analysis data of the gene conversion event in isogenic pilus variants E to F3 and F7 (9, Meyer and Haas, submitted). Numbers on *top* of the gene copies refer to codon positions of constant and variable regions in the pilus gene (12)

other hand, all SC's extend to the distal carboxy terminal end and some have a stop codon at the correct site (unpublished).

Comparing the DNA sequences, we find that most of the silent copies contain variant sequences. Some regions in each copy, however, are strictly conserved. In this way each variable sequence of a given copy is flanked by conserved regions. We designated the unit of one variable region and two flanking constant regions as minicassette (mc). They seem to represent the basic units in the reassortment of a pilin expression gene.

Conversion of Minicassettes in the Expression Locus

To determine the significance of the silent gene repertoire, we synthesized an oligonucleotide probe which was complementary to minicassette mc2 from silent copies 1 and 3 (SC1/3) in the *pil*S1 locus of MS11 variant A. This probe did not hybridize to any other gene copies in the parental MS11 genome and is therefore specific for the silent locus *pil*S1. Using this DNA probe, we analyzed spontaneous pilus variants derived from the parental MS11 strain by Southern blot hybridization. In some of these variants, an additional hybridization in the expression gene could be found while the hybridization signal with *pil*S1 was unchanged. Detailed investigation of this finding showed that the expression gene (*pil*E1) had acquired sequences from the *pil*S1 locus in a way that is reminiscent of a typical gene conversion process.

Sequence comparisons between *pil*S1(SC1/3) and *pil*E1(EC) of the parental variant and a progeny variant showed that not only the probe-analogous minicassette, but also two additional minicassettes had been transferred from *pil*S1(SC1) to the expression gene of the progeny. Due to this event the original minicassettes (mc1-3) in the expression locus were replaced, while minicassettes mc4-6 were maintained. No reciprocal changes were detectable in the silent donor locus (Fig. 2). Thus, we observed an unidirectional recombinant event that only affected the expression site. The original sequences were apparently lost from the expression locus but they were found to be present at least a second time in another silent locus of the genome and were thus preserved.

Following this initial finding, gene conversion was detected to occur also between other silent pilin loci and the expression gene(s) (Haas and Meyer, submitted). In all cases we have studied, gene conversion involved one or several complete minicassettes of the pilin genes. The endpoints of recombination were always located in the conserved flanking sequences of the minicassettes. This lead us to propose the minicassette model, which suggests minicassettes as the basic units transferred by this mechanism. The essence of this model is that gene conversion requires flanking homology at the regions to be transferred from a donor to an acceptor gene. This assumption was also found to be true for gene conversion between the trypanosome VSG genes, which seems always to take place between sites of homology of a basic and the expression-linked copy (13).

The pilin molecules expressed from genes that have undergone conversion of minicassettes form pili with altered functional and structural characteristics. Interestingly, certain pilus variants produce pilin molecules that will not be assembled into pili structures. Most likely, this incompatibility depends on the arrangement of minicassettes in the expression gene (see also (14)). Such nonpiliated variants can revert to P^+ after a second conversion of the expression gene(s) by other minicassettes from silent gene copies. The nonpiliated variants still produce pilin in normal amounts. This type of pilin (P^s), however, becomes further processed and truncated at its amino-terminal end (Haas and Meyer, submitted).

Genomic Rearrangements Associated with Pilin Genes

While P^s pilin seems to be the result of gene conversion in the expression gene, we have observed, although less frequently, other types of sequence rearrangements which are associated with clearly detectable size changes of the expression loci. Such coarse genomic rearrangements seem to fall into at least three different categories: (1) Deletion in the expression locus involving promoter and 5'-coding sequences of the expression gene (15). This type of rearrangement occurs with low frequencies and leads to a complete loss of pilus expression on both transcriptional and translational level. The deletion formation seems to be reversible only if a second intact expression copy is maintained in the genome (Haas and Meyer, unpublished; Sparling, unpublished). (2) Another type of rearrangement, recently detected (Haas and Meyer, unpublished), results in a physical translocation of the expression site presumably to a previously silent locus in the genome. Strikingly, this type of rearrangement does not cause detectable changes in the standard *Cla*I Southern hybridization pattern, while it does reveal changes when the genomic DNA is treated with other restriction enzymes. The switch of expression sites is rarely observed and may be associated with an antigenic switch in pilus expression. (3) Yet another expression site associated rearrangement occurs with high frequency. This type of recombination produces pilin sequence insertions in the expression gene (see Fig. 3). The sequence insertions are in-frame, thus giving rise to translational pilin fusions which are increased in size

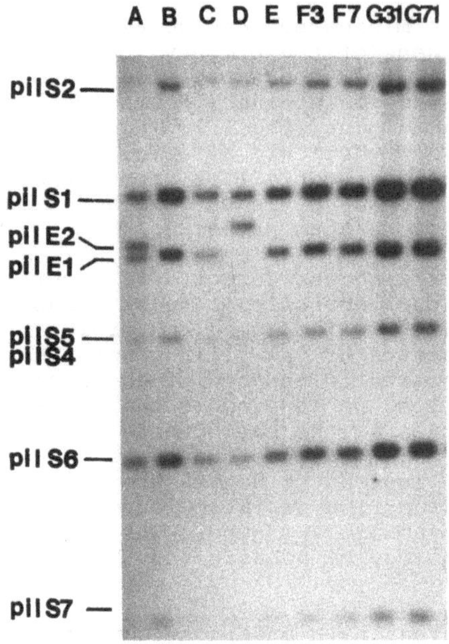

Fig. 3. Southern blot of chromosomal DNA from isogenic pilus variants (*A* to *G71*; Haas and Meyer submitted) using a constant pilus gene probe. Chromosomal DNA samples were digested with *Cla*I, run on a 0.8% agarose gel and blotted onto nitrocellulose. The 5'-labeled oligonucleotide (GC3: 5'-AACCGGCTGTCCGCAGAA-3') used for hybridization corresponded to the codon positions 120-125 of the pilus gene. Individual *pil* loci are indicated. Variants *B* to *G* have a deleted *pil*E2 locus. No size changes are detectable for the *pil*E1 locus, with the exception of variant *D* which carries an insertion of about 500 bp in its expression gene

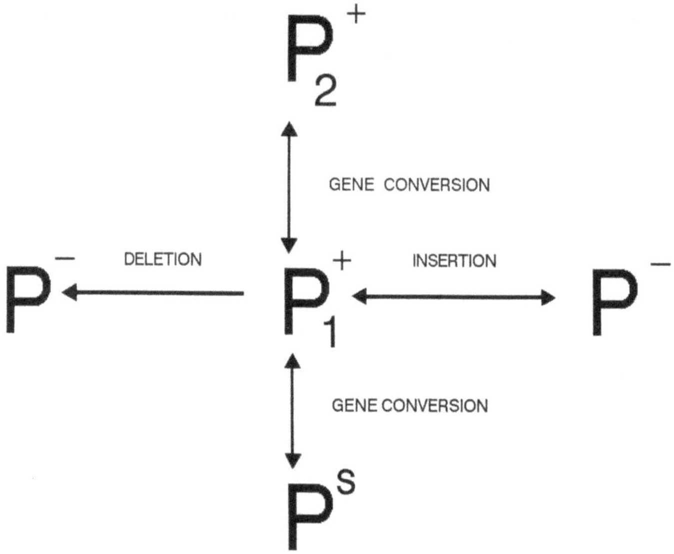

Fig. 4. The piliation phases of the gonococcus. Antigenic variation (e.g., from P_1^+ to P_2^+) frequently results from gene conversion. Gene conversion also causes cells to enter the P^S (or intermediate $P^{S/+}$) phase (Haas and Meyer submitted). The effect of gene conversion is the reassortment of variant minicassettes in the expression gene. Such reassortment can also be caused by other recombination events, such as the reversible insertion of pilin sequences in the expression gene (variant D; unpublished data). This phase transition, however, involves a P^- state significantly different from the P^S phase. By deletion of both pilus expression genes the cell seems to be blocked in a nonrevertible P^--state

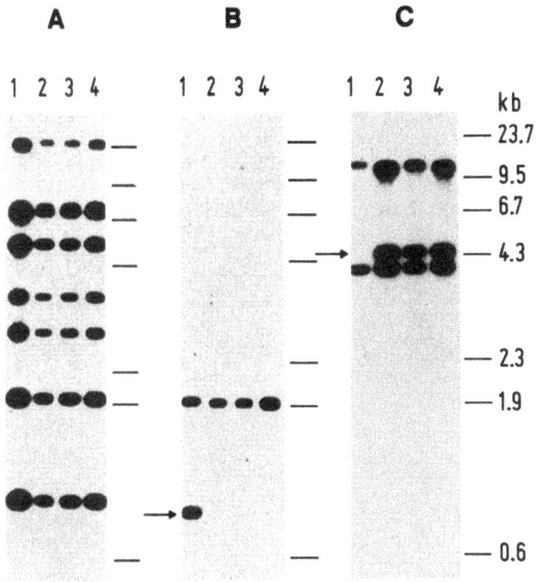

Fig. 5. Genomic blotting of chromosomal DNA with *opa* specific gene probes. Each blot (A, B and C) shows four lanes with chromosomal DNA from the parental MS11 variant VO (lanes *1*), and its opacity variants T18 (lanes *2*), V18 (lanes *3*) and V28 (lanes *4*). The DNA samples were digested with *Cla*I (A and B) or with *Sma*I (C). Genomic blots were probed with 5'-labeled oligonucleotides, corresponding to the CR sequence common to all *opa* loci (A), HV_b region in *opa*E1 of variant VO (b) and HV_b region in *opa*E1 of the progeny variant V28 (C). *Arrows* mark the *opa*E1-specific fragments that are subject to gene conversion

and therefore are not suited to become assembled into intact pili. The resulting P^- state is reversible to the P^+ state and often causes changes in the order of minicassettes in the expression gene (Haas and Meyer, unpublished). The scheme shown in Fig. 4 summarizes the most frequent changes which pilin expression genes are subjected to.

The *opa/opr* Gene Family of Pathogenic *Neisseriae*

The organization of the opacity genes in *N. gonorrhoeae* differs from the pilus genes, as the latter gene family is composed of complete genes designated as *opa*E. There is no evidence that *opa* genes are clustered in the genome as shown for the major pilin loci. Some of the *opa* loci however are closely linked to *pil* loci, in one case to the *pil*E1 locus. This *pil/opa* gene linkage exhibits conserved structural characteristics such as the conserved distance and downstream location of the *opa* loci relative to the *pil* loci. The significance of the *pil/opa* linkage is not clearly understood. It might however suggest a role in some yet unknown coordinated mechanisms of expression and/or recombination control of the two gene families (7).

As determined by sequence comparison (Stern et al., in preparation) most of the *opa* genes present in a single genome contain variant structural sequences. Similar to the pilus gene arrangement these sequence variations are clustered in three major hypervariable domains: HV_a, HV_b, and HV_c (16). According to our computer predictions, the variable domains are exposed outside the gonococcal cell surface (17).

Recently we detected the presence of genes related to the gonococcal *opa* genes in *N. meningitidis* and *N. lactamica* (16). DNA sequence analysis of such Op-related gene sequences *(opr)* reveals a strong homology to the constant regions found in the gonococcal *opa*E genes. Like the *opa*E genes, all *opr* genes seem to be complete structural entities. Together they form a large interspecies family with related structural characteristics.

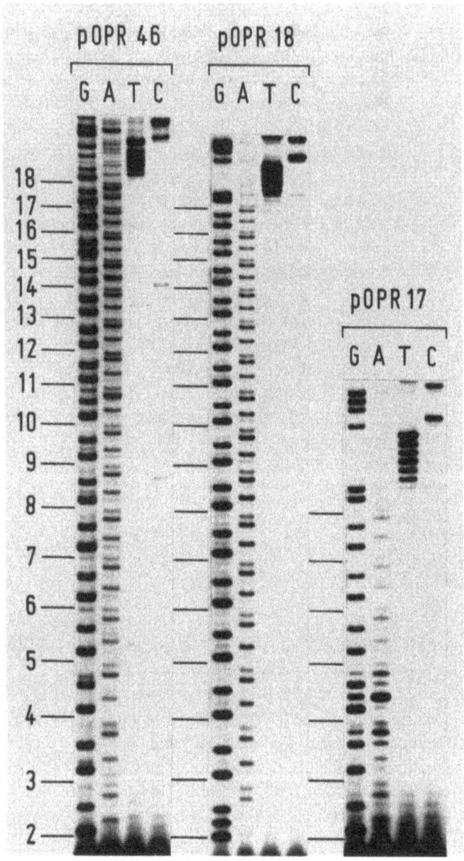

Fig. 6. DNA sequences of the coding repeat (CR) of pOPR17, pOPR18 and pOPR46 (10). The sequencing experiments were performed by a modified "supercoil" sequencing method (18) using a synthetic primer complementary to the downstream region of the CR-sequence (10). The number of repeats are indicated at the *left borders* of the sequencing gels

Individual members of the *opa* gene family were observed to exchange variant sequences among each other. This was shown by Southern hybridization using defined oligonucleotide probes of the HV_b domains of an individual *opa* locus, *opa*E1, found as the smallest band in a typical *Cla*I-restriction pattern (Fig. 5A). The HV_b probe of the parental gonococcal variant reveals in the *opa*E1 locus a corresponding sequence which apparently is lost in the progeny variants (Fig. 5B). Vice versa, the progeny variants acquire another HV_b sequence in this particular locus originating from another *opa*E locus, as specified with the corresponding sequence as probe. Clearly there are no size changes associated with this recombination. As for the pilin genes, we can explain this phenomenon best with the process of gene conversion.

Translational Control Based on the Coding Repeat Sequence

The observation that *opa* genes interact with each other by means of gene conversion does not explain why at a time none, or only one or two, *opa* genes are expressed in a distinct gonococcal variant. We observed that due to their promoters all *opa/opr* genes are constitutively transcribed (10,16). Not all of the *opa/opr* transcripts in a cell, however, are successfully translated to a functional protein. The clue to this striking expression regulation relies in a peculiar sequence found in all *opa/opr* genes, i.e., the so-called coding repeat (CR)

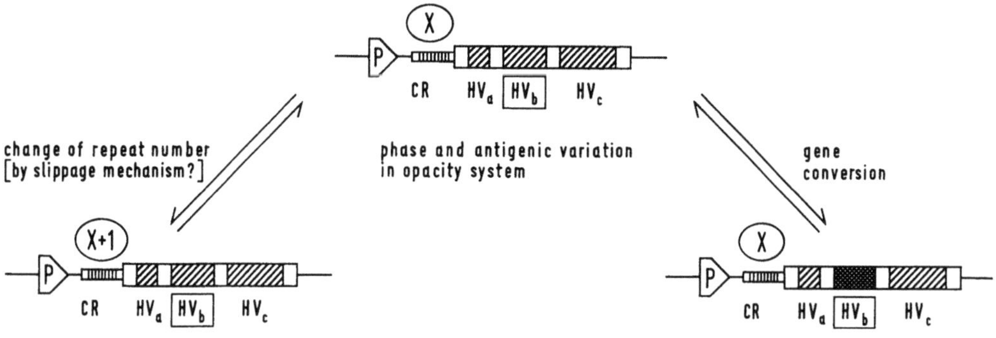

Fig. 7. Gene conversion and CR variation as the basis for opacity phase and antigenic variation. Gene conversion leads to the replacement of hypervariable sequences by a corresponding sequence of another locus. The number of CR units (X or X+1) determines whether one and which one of the *opa* genes is translationally activated. CR variation is likely to be caused by a slippage process in association with DNA replication and seems to occur more frequently than gene conversion

that codes for the hydrophobic portion of the *opa/opr* leader peptide. Examples of the CR sequences consisting of CTCTT pentameric repeat units are shown in Fig. 6 for a number of different *opa* loci of a single gonococcal variant. Clearly, the loci shown have different numbers of repeat units, ranging from 7 up to 28 (in Fig. 6 examples for 8, 17 and 18 repeat units are shown). It was therefore suggestive that the number of CR pentamer units might be the crucial means by which the reading frame, and therefore translation, of individual *opa/opr* loci is controlled.

To prove this striking hypothesis, we designed a specific primer extension experiment that should tell us the number of CR units in each of the different *opa/opr* transcripts of a single cell. The primer used was complementary to the region proximal to the 3'-end of the CR sequence in *opa/opr* transcripts and was extended with reverse transcriptase in the presence of only two nucleoside triphosphates, dATP and dGTP. These triphosphates were sufficient to produce the complementary strand to the *opa/opr* RNA's, but caused an abrupt stop because the primer could not be extended beyond the 5'-end of the CR sequence. Accordingly, the extended primers formed distinct bands in a polyacrylamide gel corresponding to the number of CR units in the respective transcripts (Fig. 6).

This style of experiment was performed with a series of isogenic opacity variants of *N. gonorrhoeae N. meningitides*, and *N. lactamica*, which were positive (O^+) or negative (O^-) in the production of opacity or opacity-related proteins. We could demonstrate by this method that *N. gonorrhoeae* harbors at least 8, probably 12, *opa* loci (K. Muralidharan, unpublished), *N. meningitidis* harbors four *opr* loci and *N. lactamica* two, which are all constitutively transcribed regardless of their state of expression. More importantly, we could show that changes in the expression of *opa/opr* are inherently associated with changes in the number of CR units in distinct transcripts. Precise evaluation on the nucleotide sequence level confirmed that changes in the repeat number of individual gene loci affect the reading frame of these genes and are ultimately responsible for both phase and antigenic transitions in the pathogenic *Neisseriae* (10,16,19) (see Fig. 7).

Acknowledgment. We should like to thank K. Lamberty for the art work and T. Gretzinger for photography. Our studies were supported by a grant from the Deutsche Forschungsgemeinschaft (Me705/2-3).

References

1. Stephens DS, McGhee ZA (1983) Association of virulence of *Neisseria meningitidis* with transparent colony type and low-molecular-weight outer membrane proteins. J Infect Dis 147:282-292
2. Poolman JT, Hopman CTP, Zanen HC (1985) Colony variants of *Neisseria meningitidis* strain 2996 (B.2b:P 1.2): Influence of class 5 outer membrane proteins and Lipopolysaccharides. J Med Microbiol 19:203-209
3. Kristiansen BE, Sorensen B, Simonson T, Spanne O, Lund V, Bjorvata B (1984) Isolates of *Neisseria meningitidis* from different sites in the same patient: Phenotypic and genomic studies, with special reference to adherence, piliation, and DNA restriction endonuclease pattern. J Infect Dis 150:389-396
4. Blake SM, Gotschlich EC (1983) Gonococcal membrane proteins: speculation on their role in pathogenesis. Prog Allergy 33:298-313
5. Schoolnik GK, Tai JY, Gotschlich EC (1983) A pilus peptide vaccine for prevention of Gonorrhea. Prog Allergy 33:314-331
6. Hagblom P, Segal E, Billyard E, So M (1985) Intragenic recombination leads to pilus antigenic variation in *Neisseria gonorrhoeae*. Nature (London) 315:156-158
7. Stern A, Nickel P, Meyer TF, So M (1984) Opacity determinants of *Neisseria gonorrhoeae*: gene expression and chromosomal linkage to the gonococcal pilus gene. Cell 37:447-456
8. Meyer TF, Mlawer N, So M (1982) Pilus expression in *Neisseria gonorrhoeae* involves chromosomal rearrangement. Cell 30:45-52
9. Haas R, Meyer TF (1986) The repertoire of silent pilus genes in *Neisseria gonorrhoeae*: evidence for gene conversion. Cell 44:107-115
10. Stern A, Brown M, Nickel P, Meyer TF (1986) Opacity genes in *Neisseria gonorrhoeae*: control of phase and antigenic variation. Cell 47:61-71
11. Swanson J, Bergström S, Barrera O, Robbins K, Corwin D (1985) Pilus-gonococcal variants: evidence for multiple forms of piliation control. J Exp Med 162:729-744
12. Meyer TF, Billyard E, Haas R, Störzbach S, So M (1984) Pilus gene of *Neisseria gonorrheae*: chromosomal organization and DNA sequence. Proc Natl Acad Sci USA 81:6110-6114
13. Pays E, Delauw M-F, Van Assel S, Laurent M, Vervoort T, Van Meirvenne N, Steinert M (1983) Modifications of a Trypanosoma b. brucei antigen gene repertoire by different DNA recombinational mechanisms. Cell 35:721-731
14. Bergström S, Robbins K, Koomey JM, Swanson J (1986) Pilation control mechanisms in *Neisseria gonorrhoeae*. Proc Natl Acad Sci USA 83:3890-3894
15. Segal E, Billyard E, So M, Störzbach S, Meyer TF (1985) Role of chromosomal rearrangement in *N. gonorrhoeae* pilus phase variation. Cell 40:293-300
16. Stern A, Meyer TF (1987) Common mechanism controlling phase and antigenic variation in pathogenic Neisseriae. Mol Microbiol 1:5-12
17. Meyer TF, Fiedler H, Frosch M, Haas R, Veit S, Jähnig F, Muralidharan K, Stern A (1987) Variable and conserved proteins on the surface of pathogenic Neisseriae. In: Bacterial vaccines and mucosal immunity. Annali Sclavo, Siena, in press
18. Wallace RB, Johnson MJ, Suggs SV, Miyoshi K, Bhatt R, Itakura K (1981) A set of synthetic oligodeoxyribonucleotide primers for DNA sequencing in the plasmid vector pBR322. Gene 16:21-26
19. Black WG, Schwalbe RS, Nachamrin I, Cannon JG (1984) Characterization of *Neisseria gonorrhoeae* protein II phase variation by use of monoclonal anitbodies. Infect Immun 45:453-457

Molecular and Antigenic Architecture of the Antiphagocytic Group A Streptococcal M Protein

V. A. Fischetti[1], K. F. Jones[1], S. Hollingshead[2], and J. R. Scott[2]

Introduction

The ability of the group A streptococcus to cause human infection is primarily attributable to the M protein, a molecule located on the surface of the cell wall. This protein enables the organism to resist phagocytic attack by human neutrophils, thus allowing the streptococcus to persist in the infected tissue until such time as antibodies are produced against the M molecule (1). Because type-specific antibodies to the M protein lead to the effective clearance of the invading streptococcus, the ability of the organism to vary the antigenic structure of the M molecule has probably been responsible for the survival of the streptococcus in nature (1,2). Lancefield showed that, though delayed in their appearance, once formed, type-specific opsonic antibodies persist in the host for more than 30 years (2). To date, more than 80 serologically distinct M molecules have been identified in addition to a large number of clinical isolates for which typing sera are not yet available ((3), WHO meeting, Tokyo, Japan 1985).

The presence of a large number of different antigenic serotypes of M protein is indicative of a mechanism by which the streptococcus is able to alter this molecule. Our structural, immunological, and genetic studies have allowed us to better understand the mechanism by which antigenic variation may be occurring in the streptococcus.

Using DNA isolated from the M6 streptococcal strain D471 which was partially digested with *Sau3a* and ligated into the *BamHI* site of a cosmid vector, the M6 gene was successfully cloned into *E. coli* (4). By structural, functional and immunological techniques, the protein produced in this new host appeared to be the same as that produced in the streptococcus (4,5). In order to ask specific questions regarding the functional regions of the M protein, a system was devised whereby the M protein gene *(emm)* could be moved from *E. coli* back to a group A streptococcal strain (6) that we had found to be deleted for all *emm*-homologous DNA (7). Unlike the parental M- recipient, a transconjugant strain that acquired the *emm*6 gene expressed M6 protein on its surface and survived in human blood. In addition, this strain absorbed M6 opsonic antibodies from hyperimmune rabbit serum and generated opsonic antibodies when injected into rabbits. Thus, the plasmid containing the *emm*6 gene appears to confer on the M- organism all the traits associated with M protein (1).

Structural Characteristics

In our earlier studies on the M proteins from three different M serotypes, it became apparent that except for the N-terminal 10 - 15 amino

[1]The Rockefeller University, 1230 York Avenue, New York, NY 10021, USA

[2]Department of Microbiology, Emory University, Atlanta, GA 30332, USA

38. Colloquium Mosbach 1987
Molecular Basis of Viral and Microbial Pathogenesis

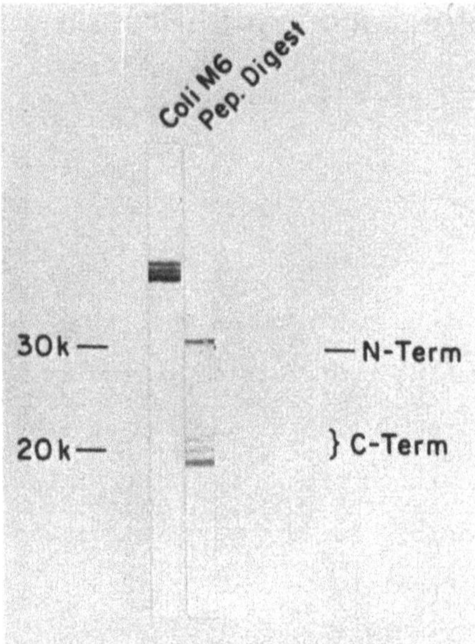

Fig. 1. Western blot analysis of M6 protein produced in *E. coli* containing the gene for M6 protein and a pepsin digest of the protein. The purified protein was cleaved with pepsin at pH 5.8 and the digest was run along with the undigested molecule on SDS-PAGE. After western blotting into nitrocellulose, the M protein was resolved with polyclonal antiserum to the M6 protein. To determine the position of the N- and C-terminal fragments, a separate blot was reacted with either monoclonal 10A11 (specific for the N-terminal fragment) or 10B6 (specific for the C-terminal fragment (14). As can be seen, the triple band, characteristic of the native molecule is associated with the C-terminal fragment while the N-terminal fragment does not exhibit this pattern

acids, these molecules all have a seven-residue periodicity in the placement of hydrophobic residues within the amino acid sequence (8). This type of periodicity is characteristic of molecules whose tertiary structure is either completely or partially coiled coil (e.g., tropomyosin, myosin, fibrinogen) (8). Ultracentrifugation studies revealed that the M proteins exist as stable dimers and circular dichroism spectra indicated that about 70% of the M6 molecule is alpha-helical and thus likely to form a coiled-coil structure (9). Fiber X-ray diffraction data further substantiated this structure (9). The seven-residue periodicity of hydrophobic amino acids has subsequently been found in sequences for molecules of other M serotypes, including the partial sequence of M24 (10), the complete sequence of the N-terminal half of M5 released by partial pepsin digestion (11), and that which is predicted from the complete DNA sequence of M6 ((12), Fischetti et al., in press).

The gene for M6 was successfully cloned into *E. coli* (4). The M molecule produced in this organism accumulated in the periplasmic space and was not found either on the surface of the organism or in the culture medium (5). The molecule produced consisted of three protein bands of apparent molecular weight 55,000, 57,000, and 59,000. Since the N-terminal sequence of this preparation was homogeneous and agreed with the DNA sequence, the reason for the three bands probably resides in another part of the molecule. Cleavage of the protein with pepsin at pH 5.8 (13), results in a homogeneous protein band, representing the N-terminal half, and three bands representing the C-terminal half of the molecule. These fragments were identified by specific monoclonal antibodies directed against the N- and C-terminal regions (Fig. 1) ((14), unpublished data). Thus, the cause for the three-band pattern is localized within the C-terminal segment of the molecule.

The complete nucleotide sequence of the M6 gene has revealed for the first time the characteristics of the amino acid sequence of an entire M molecule (12). DNA sequence analysis of the C-terminal region of the M6 molecule revealed a stretch of 20 essentially hydrophobic amino acids which could act as a membrane anchor, and of an adjacent proline-

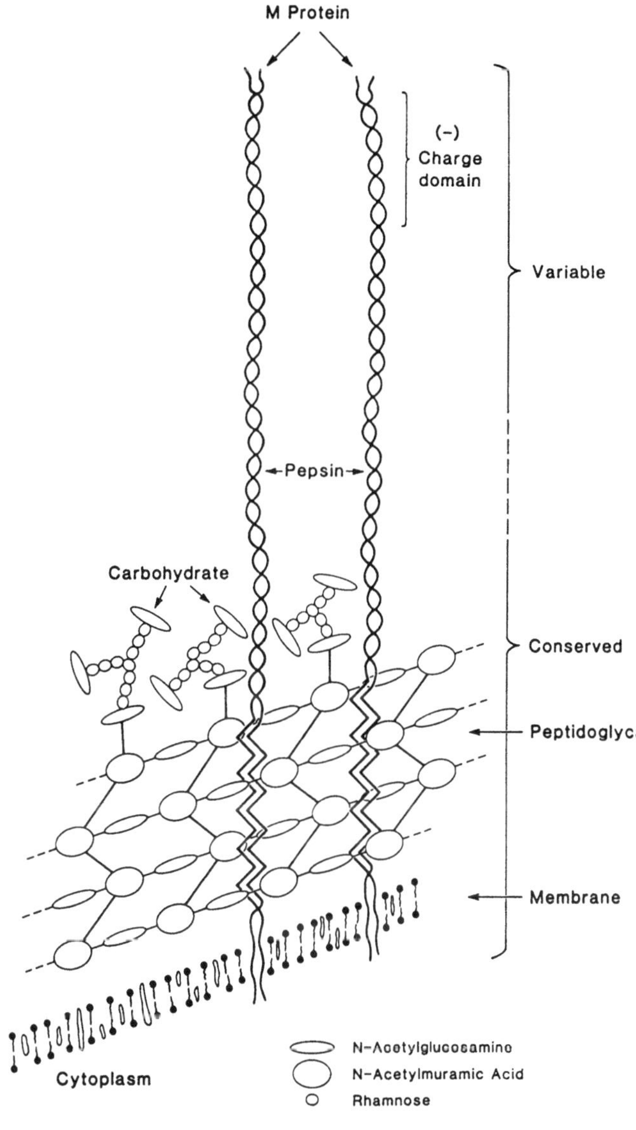

Fig. 2. Schematic representation of the coiled-coil M protein molecule on the streptococcal cell wall. Pepsin represents the region of the molecule cleaved by pepsin at pH 5.8. A negative charge domain is located at the N-terminal region based on sequence data on the M5 (23) and M6 proteins (Fischetti et al., in preparation). Variable and conserved regions are defined on the basis of DNA hybridization studies between other M serotypes (26) and reactivity with monoclonal antibodies (14). From current data, the molecule appears to be held in the cell by both a membrane anchor at the C-terminus and an adjacent proline-glycine-rich wall stabilization domain likely to be emedded within the peptidoglycan (12). Carbohydrate represents the group A specific carbohydrate. Model is not drawn to scale to emphasize the wall anchor region

and glycine-rich region (12). The structure produced by the helix-disrupting proline and glycine residues may cause this region of the M molecule to assume an irregular beta-sheet conformation, allowing it to intercalate within the cross-linked region of the peptidoglycan, thus stabilizing the molecule within the cell wall. Evidence from one of our laboratories indicates that the M molecule is not covalently linked to the cell wall matrix (Pancholi and Fischetti, submitted). Therefore, we speculate that since the M protein is synthesized at the septum (15), where the cell wall is least cross-linked (16), the membrane anchor may be required to prevent release of the native protein during its maturation. The peptidoglycan is then cross-linked around the proline-glycine region, thus stabilizing the M molecule in the cell wall. Sequences highly homologous to the C-terminal region are also found at the C-terminal end of staphylococcal protein A (12) and the

IgG binding protein of group G streptococci (17), both of which are surface proteins of other gram-positive organisms. This type of structural domain for surface proteins may allow them to be anchored in the cell wall and thus, may be a common feature of such proteins in gram-positive organisms.

In summary, the present data are consistent with the idea that the M protein is a coiled-coil fibrillar structure composed of two alpha-helical protein chains extending about 600 Å from the surface of the streptococcus (9) and anchored at both the cell wall and membrane ((12, 18), Fischetti et al., in press) (Fig. 2).

Immunological Properties

While a great deal of immunological information is available about the M protein molecule, the majority of these data is derived from a fragment of the native molecule released from the streptococcus by partial pepsin digestion (PepM) (10,11,19,20). This fragment comprises approximately the N-terminal half of the native M protein (9,20) and contains the determinants necessary to induce type specific opsonic antibodies when injected into rabbits.

When whole streptococci were used to produce antibodies to the M6 protein, cross-reactions were observed between these sera and certain heterologous M proteins (21). Peptide map analysis of the PepM protein from two cross-reactive types revealed eight peptides in common, while non-cross-reactive types had only two common peptides (22). These early results suggested that some M proteins were more closely related than others. However, in general, both immunological and peptide studies indicated that the N-terminal half of the M proteins from different serotypes differed significantly. Despite apparent amino acid sequence variation within this region of the M molecule, a common antiphagocytic activity persists.

A common feature of the amino acid sequence of the N-terminal domain among the M5 and M6 proteins is the presence of a significantly higher concentration of negatively charged amino acids, predominantly glutamic acid (23), Fischetti et al., in press). Since human neutrophils have a net negative surface charge, the negative charge on the M protein may prevent or retard contact between the streptococcus and the phagocyte through electrostatic repulsion. This may play a role in the antiphagocytic effect of the molecule. In support of this, it was found that human opsonic antibodies to M6 streptococci are more basic than nonopsonic M antibodies, suggesting that the negatively charged domain of the M protein is important for the generation of opsonic antibodies (24).

Killed M^+ streptococci were used to determine whether resistance to phagocytosis is the result of the ability of M protein to impair the function of the invading phagocytic cell. The presence of killed phagocytosis-resistant M^+ streptococci, in a nearly 1000-fold excess, did not interfere with the phagocytosis of live M- streptococci (25). Thus, in this mixture, the phagocyte was able to seek out and effectively destroy the M- organisms, while ignoring the M^+ cells. These results also tend to support the view that a repulsive charge may play a role in the antiphagocytic activity of the molecule. However, if there is a direct inhibitory effect of the M protein on the phagocyte it may be confined to the area of the phagocyte membrane in direct contact with or in close proximity to the M protein.

DNA hybridization studies were used to determine whether sequences within the M6 molecule are common to regions within M protein from different serotypes. Using a labeled probe consisting of nearly the complete M6 gene, we found hybridization with DNA from all 56 different M serotypes tested, in addition to four strains for which typing sera is not yet available (7). Using smaller intragenic probes, we found that the C-terminal region of the gene is conserved among strains of different serotypes, while the N-terminal half is more variable among these strains (26).

Monoclonal antibodies prepared against the native M6 protein were used to define antigenic determinants on the M molecule. Based on the reactivity of three different antibodies in a colony blot assay, the N-terminal, pepsin-derived fragment contained determinants shared by 10/56 different serotypes, while more centrally located determinants within the C-terminal half of the molecule were each found to be conserved among 30/56 different serotypes (14,27).

The results thus far suggest that the N-terminal region, which is exposed to the immune system, is variable with regard to amino acid sequence but conserved with respect to net charge. The amino acid sequence of the region proximal to the cell wall is more conserved among different M proteins. From immunological and sequence comparisons of the pepsin fragment from M5 and M6 proteins, the N-terminal half of the native molecule may be further divided into a variable domain (near the center of the native molecule) and a hypervariable domain (N-terminal 25%) (Fig. 2) (23).

Size Variation and Its Relation to Antigenic Variation

In addition to the characteristic antigenic variation of the M molecules, we recently discovered that there is variation of as much as 40,000 in the apparent molecular weight of M proteins isolated from strains of the same or different streptococcal serotypes (28). This may be explained by our finding that approximately the N-terminal 2/3 of the mature M molecule is composed of three repeat regions (12). Region A is composed of five direct tandem repeats of 14 amino acids each, region B of five tandem repeats of 25 amino acids each and region C of two non-tandem direct repeats of about 27 amino acids each (Fig. 3). Extensive amino acid sequence repeats have been reported to occur in the PepM24 fragment (10), and they have been also found, but are less common, in the PepM5 fragment (20). Such long reiterated DNA sequences are likely to serve as substrates for homologous recombination, which should generate deletions and duplications within the M protein gene leading to the production of M proteins of different sizes.

To test this hypothesis, we needed to sequence the M protein gene from a set of clonally related strains. Thus, it was necessary to isolate M protein size mutants in the laboratory so that the parentage of the strain would be known. A Western blot system was devised such that M protein size mutants could be detected and isolated if they constitute at least 5% of the tested culture. From type 6 strain D471, about 1 in 2000 CFU was found to produce a smaller size M protein (28).

Analysis of the DNA sequence of the M6 genes of four spontaneous M protein deletion mutants isolated in our laboratory confirmed that these mutants arise by homologous recombination between reiterated DNA sequences within the structural gene encoding M protein (29). By sequence analysis, each of the mutants with a different size M protein was found to have a deletion of one or more of the intragenic repeat blocks.

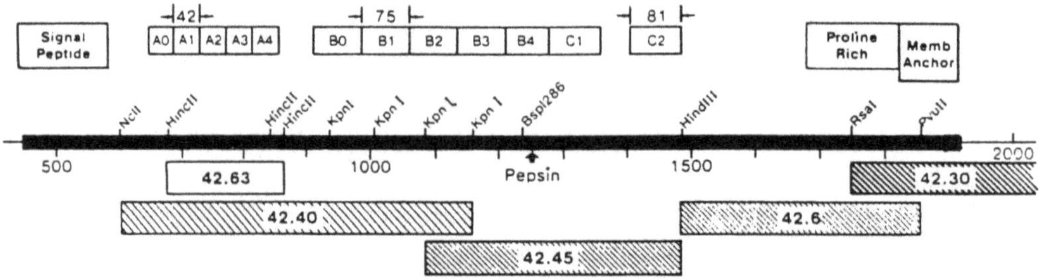

Fig. 3. Major features of the DNA sequence of the M6 gene. The sequence encoding the mature M6 protein is indicated by the *bold black* box. Relevant restriction enzyme sites are indicated above this line and the scale below it is in kb. The predominant site at which pepsin cleaves the M6 protein is indicated by the *bold arrow*. The boxes in the *top row* indicate the location of the proline-glycine-rich region and membrane anchor as deduced from the DNA sequence (12). The three regions of reiterated DNA sequence are delimited as boxes *A*,*B*, and *C* and the number of base pairs in each is indicated above a box in each region. Below the coding sequence, the boxes indicate the probes used in DNA hybridization experiments and the degree of shading in these boxes represents the proportion of tested group A streptococcal strains that show homology with each probe (26). The *darkest shading* probe (42.30) indicates homology with all strains tested, and the *unshaded box* (probe 42.63) indicates no homology with any strains except the one from which the probe was derived

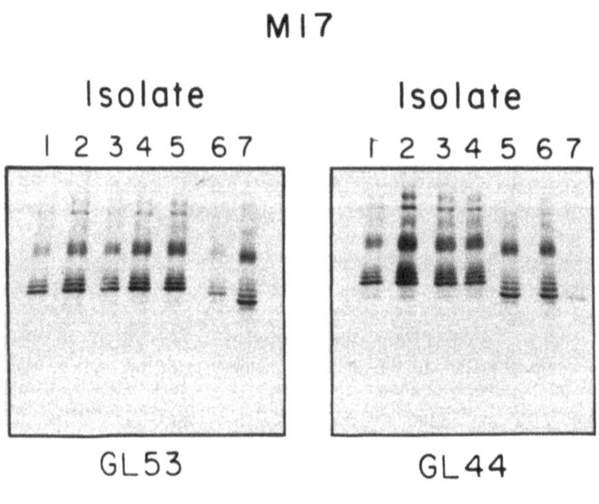

Fig. 4. Western blot of lysin extracts illustrating the M protein produced by serial streptococcal isolates from two patients with pharyngitis. Lanes represent weekly streptococcal isolates from the time the patient presented with pharyngitis (week 1). Seven weekly isolates of M17 streptococci were obtained from each patient (GL53 and GL44) without treatment during an epidemic in 1942. The strains were grown to comparable optical densities and the M protein extracted with phage lysin, which solubilizes the cell wall, releasing the M protein. After SDS-PAGE and Western blot, the nitrocellulose was probed with a monoclonal antibody reactive with M17 protein. The blots represent the typical multiple banding pattern of M protein extracted by this method (18). It may be seen that the size of the M protein produced by the streptococci was the same for patients GL53 and GL44 to the 5th and 4th weekly isolate, respectively. After that, the streptococci isolated from the throat of these individuals synthesized a smaller M protein (though still typed as M17). In the case of patient GL53, a streptococcus producing an even smaller size M protein was isolated in the 7th week. While the same size, the amount of M protein produced by the 7th isolate from patient GL44 was less than previous isolates. A reduction in the quantity of M produced is also seen in the 6th isolate of patient GL53. This may be an example of the observed reduction in M protein produced by streptococci during the carrier state

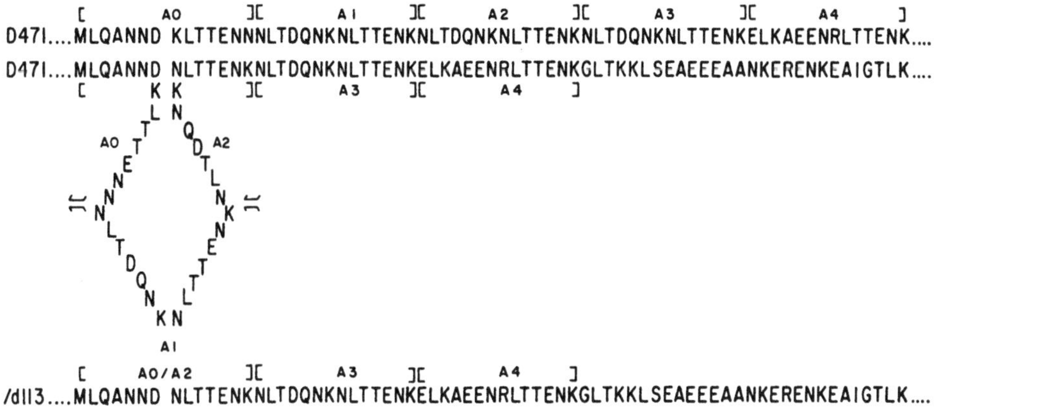

Fig. 5. Alignment of the amino acid sequence of the A repeat region of the M6 protein from parental strain D471 before and after looping out two A repeat blocks (28, 29). The deletion of the two A repeat blocks resulted in the sequence shown for strain D471d113 (29), which most likely arose by a homologous recombination event. This process was able to generate a new sequence for this M6 protein derivative

Normally, when beta-hemolytic streptococci are isolated from a blood agar plate either as primary isolates from clinical material or during the maintenance of a laboratory strain, a single hemolytic colony is chosen at random for further growth. Thus, in spite of their relatively frequent occurrence, mutants with variant M protein size are not likely to be isolated during routine laboratory manipulations (18). However, M protein size mutants were found in nature, both among strains of the same M type obtained serially from the same patient (18) (Fig. 4), and among strains isolated in single streptococcal outbreaks or epidemics (18,29). This observation of size variation among clinical isolates indicates that the size mutant constituted the major organism in the streptococcal population at the time of isolation. This suggests that the size mutant may have some selective advantage under clinical conditions.

Based on DNA sequence analysis, the size variation of at least six clinical M6 isolates appeared to be the result of homologous recombination within the repeat sequences (29) as was found with the mutants isolated in the laboratory. The natural isolates were in two groups: serial isolates from the same patient, and isolates from a single outbreak. Sequence analysis of the M gene also confirmed that the strains from each group studied appeared to be clonally related and not the result of separate aquisitions of unrelated M6 strains (29).

Because M proteins of all the different serotypes tested appear to be related (7,26), they probably arose from a common ancestor or directly from each other. We found that streptococcal strain D471, from which the M gene was cloned, has no extragenic regions of significant homology with the M protein gene (7), so antigenic variation of this protein is not apt to occur by extragenic recombination as found in Neisseria (30) and Borrelia (31). On the other hand, it seems likely that intragenic homologous recombination events contribute to the antigenic diversity among M serotypes (29). When changes in the amino acid sequence arise as a result of mutation or recombination between inexact repeats, they may be amplified by homologous recombination between intragenic repeated sequences. If these sequences are involved in antigenic determinants, they may be amplified or in some cases deleted by such recombination events (29). Thus, intragenic homologous recombination would be expected to accelerate the rate at which alterations accumulate in the M protein gene.

In both the A and B repeat regions of the M6 sequence, the terminal repeat blocks diverge slightly from the consensus sequence of the three internal blocks. Homologous recombination between one of these terminal blocks and any other central block might generate a new sequence. Such a recombination event appeared to have occurred in the A repeat region of a deletion mutant isolated in the laboratory from strain D471 (Fig. 5). If such a new sequence occurred in a biologically important region of the molecule, i.e., an opsonogenic immunodeterminant or a region necessary for colonization, it might lead to a change in the activity of that region. Both A and B repeat regions contain antigenic epitopes (27), which we might expect to be immunodominant since they constitute a large proportion of the exposed M molecule. Studies are in progress that will help clarify this question. However, it seems apparent that antigenic diversity may arise by intragenic homologous recombination and that changes that result from this event might alter the antigenic structure of the molecule. An antigenic alteration that would be important to the streptococcus is one that leads to its escape from recognition by opsonic antibodies produced in the host.

Thus, through immunological pressure, the antigenic structure of the M protein is continuously evolving by the selection of organisms producing variant M molecules. This process is likely to result in a continuum of antigenically unique M molecules among the streptococci in the environment.

Acknowledgments. We wish to thank Mary Windels, Clara Eastby, Christina Gotschlich, Lynn Malone, William Pulliam, Sharon Coyle, Patty Guenthner, Andrew David, and Jane Rhatigan for their excellent technical assistance during the course of some of these studies. These studies were supported in part by NIH Grants AI11822 (to VAF) and AI20723 (to VAF and JRS).

References

1. Lancefield RC (1982) Current knowledge of the type specific M antigens of group A streptococci. J Immunol 89:307-313
2. Lancefield RC (1959) Persistance of type specific antibodies in man following infection with group A streptococci. J Exp Med 110:271-292
3. Quinn RW, Vander Zwaag R, Lowry PN (1985) Acquisition of group A streptococcal M protein antibodies. Ped Inf Dis 4:374-378
4. Scott JR, Fischetti VA (1983) Expression of streptococcal M protein in *E. coli*. Science 22:758
5. Fischetti VA, Jones KF, Manjula BN, Scott JR (1984) Streptococcal M6 protein expresed in *Escherichia coli*. Localization, purification and comparison with streptococcal-derived M protein. J Exp Med 159:1083-1095
6. Scott JRP, Guenthner PC, Malone LM, Fischetti VA (1986) Conversion of an M- group A streptococcus to M+ by transfer of a plasmid containing an M6 gene. J Exp Med 164:1641-1651
7. Scott JR, Pulliam WM, Hollingshead SK, Fischetti VA (1985) Relationship of M protein genes in group A streptococci. Proc Natl Acad Sci USA 82:1822-1826
8. Manjula BN, Fischetti VA (1980) Tropomyosin-like seven residue periodicity in three immunologically distinct streptococcal M proteins and its implication for the antiphagocytic property of the molecule. J Exp Med 151:695-708
9. Phillips GN, Flicker PF, Cohen C, Manjula BN, Fischetti VA (1981) Streptococcal M protein: alpha-helical coiled-coil structure and arrangement on the cell surface. Proc Natl Acad Sci USA 78:4689-4693

10. Beachey EH, Seyer JM, Dale JB, Hasty DL (1983) Repeating covalent structure and protective immunogenicity of native and synthetic polypeptide fragments of type 24 streptococcal M protein. J Biol Chem 258:13250-13257
11. Manjula BN, Acharya AS, Mische SM, Fairwell T, Fischetti VA (1984) The complete amino acid sequence of a biologically active 197-residue fragment of M protein from type 5 group A streptococci. J Biol Chem 259:3686-3693
12. Hollingshead SK, Fischetti VA, Scott JR (1986) Complete nucleotide sequence of type 6 M protein of the group A streptococcus: repetitive structure and membrane anchor. J Biol Chem 261:1677-1686
13. Beachey EH, Campbell GL, Ofek I (1974) Peptic digestion of streptococcal M protein. II. Extraction of M antigen from group A streptococci with pepsin. Infect Immun 9:891-896
14. Jones KF, Manjula BN, Johnston KH, Hollingshead SK, Scott JR, Fischetti VA (1985) Location of variable and conserved epitopes among the multiple serotypes a streptococcal M protein. J Exp Med 161:623-628
15. Swanson J, Hsu KC, Gotschlich EC (1969) Electron microscopic studies on streptococci. I. M antigen. J Exp Med 130:1063-1091
16. Fan DA, Beckman BE (1973) Structural difference between walls from hemispheral caps and partial septa of *Bacillus subtilis*. J Bacteriol 114:790-797
17. Fahnestock SR, Alexander P, Nagel J, Filpula D (1986) Gene for an immunoglobulin-binding protein from a group G streptococcus. J Bacteriol 167:870-880
18. Fischetti VA, Jones KF, Scott JR (1985) Size variation of the M protein in group A streptococci. J Exp Med 161:1384-1401
19. Beachy EH, Seyer JM, Kang AH (1980) Primary structure of protective antigens of tpye 24 streptococcal M protein. J Biol Chem 255:6284-6289
20. Manjula BN, Mische SM, Fischetti VA (1983) Primary structure of streptococcal pep M5 protein: Absence of extensive sequence repeats. Proc Natl Acad Sci USA 80:5475-5479
21. Fischetti VA (1977) Streptococcal M protein extracted by nonionic detergent. II. Analysis of the antibody response to the multiple antigenic determinants of the M-protein molecule. J Exp Med 146:1108-1123
22. Fischetti VA (1978) Streptococcal M protein extracted by nonionic detergent. III. Correlation between immunological cross-reactions and structural similarities with implications for antiphagocytosis. J Exp Med 147:1771-1778
23. Manjula BN, Trus BL, Fischetti VA (1985) The presence of two distinct regions in the coiled-coil structure of the streptococcal Pep M5 protein: relationship to mammalian coiled-coil proteins and implications to its biological properties. Proc Natl Acad Sci USA 82:1064-1068
24. Fischetti VA (1983) Requirements for the opsonic activity of human IgG directed to type 6 group A streptococci: net basic charge and intact Fc region. J Immunol 130:896-902
25. Manjula BN, Schmidt ML, Fischetti VA (1985) Unimpaired function of human phagocytes in the presence of phagocytosis-resistant group A streptococci. Infect Immun 50:610-613
26. Scott JR, Hollingshead SK, Fischetti VA (1986) Homologous regions within M protein genes in group A streptococci. Infect Immun 52:609-612
27. Jones KF, Khan SA, Erickson BW, Hollingshead SK, Scott JR, Fischetti VA (1986) Immunochemical localization and amino acid sequence of cross-reactive epitopes within the group A streptococcal M6 protein. J Exp Med 164:1226-1238
28. Fischetti VA, Jarymowycz M, Jones KF, Scott JR (1986) Streptococcal M protein size mutants occur at high frequency within a single strain. J Exp Med 164:971-980
29. Hollingshead SK, Fischetti VA, Scott JR (1987) Size variation in group A streptococcal M protein is generated by homologous recombination between intragenic repeats. Mol Gen Genet 207:196-203
30. So M (1986) The pilus of Neisseria gonorrhoeae: phase and antigenic variation. In: Inouye (ed) Bacterial outer membranes as model systems, vol 1. Wiley, New York (in press)
31. Meier JT, Simon MI, Barbour AG (1985) Antigenic variation is associated with DNA rearrangements in a relapsing fever Borrelia. Cell 41:403-409

Bacterial Cytolysins that Form Transmembrane Pores

S. Bhakdi[1] and J. Tranum-Jensen[2]

Introduction

Many bacterial cytolysins damage mammalian target cells by inserting themselves into the lipid bilayer and thereby creating transmembrane pores. Phenomenologically, this process is akin to that of complement cytolysis mediated by C5-C9 (1-3) and to the mechanism of T-cell cytotoxicity (4). Further pore-forming proteins have been found amongst toxins from fungi (5), amoeba (6,7) and sea anemone (8). This chapter will focus on mechanisms of pore formation by three bacterial cytolysins that probably represent prototypes for the majority of other channel formers. The first section will summarize general features of pore formation. Thereafter, properties of staphylococcal α-toxin, streptolysin-O, and *E. coli* hemolysin will be discussed. Finally, secondary pathophysiological reactions elicited by the pore-forming cytolysins will be described.

General Features of Transmembrane Pore Formation

The unifying feature of all pore formers is that they are produced and released by bacteria initially as water-soluble proteins, but are able to undergo a unique and possibly always irreversible transition from a hydrophilic to an amphiphilic state upon interaction with a target lipid bilayer. Primary binding to a target membrane may require the presence of specific binder molecules, the best-documented example for this type of interaction being represented by the sulfhydryl-activated cytolysins (e.g., streptolysin-O), which probably initially bind to membrane cholesterol (9-11). Often, however, the cytolysins will be of the "receptorless" type that binds indiscriminately to lipid bilayers. Examples are the α-toxin of *S. aureus*, and *E. coli* hemolysin. The factors influencing the binding of such toxins to cell will be complex, and net binding will tend to be generally ineffective. These toxins may have more opportunity to diffuse longer distances in target tissues, with relatively small numbers of cytolytically effective molecules becoming deposited on the cells that come under attack.

Pores may be generated by insertion of protein monomers into the bilayer, such as is probably the case with *E. coli* hemolysin, or by the aggregation of monomers to form noncovalently associated oligomers, such as with α-toxin and streptolysin-O. Monomeric pores are too small to be visualized by electron microscopy, whereas several oligomeric pores have been amenable to ultrastructural characterization. In the latter cases, pores are generally seen as partially of fully circula-

[1]Institute of Medical Microbiology, University of Giessen, Schubertstraße 1, D-6300 Giessen, FRG

[2]Anatomy Institute C, University of Copenhagen, Blegdamsvej 3C, DK-2200 Copenhagen N, Denmark

38. Colloquium Mosbach 1987
Molecular Basis of Viral and Microbial Pathogenesis

rized structures. Oligomeric pores will usually be heterogeneous in size, due to variations in the number of protomers that constitute the individual channels. Wide variations are displayed by pores generated by streptolysin-O and other sulfhydryl-activated cytolysins, whereas α-toxin oligomers tend to be more homogeneous. The effective functional diameters of pores span a very large range, from approximately 1.5 nm (α-toxin) to over 35 nm (streptolysin-O). Even the smallest pores are, however, large enough to permit rapid transmembrane flux of ions and small molecules. Generally, oligomeric pores have been found to display little ion selectivity and no dependence on membrane potential. In contrast, monomeric pores created by *E. coli* hemolysin require a correct transmembrane potential and display a strong selectivity for cations over anions.

Although the primary structures of α-toxin and *E. coli* hemolysin are known (12,13), the domains that are membrane-embedded in these and other pore formers have yet to be identified. No details are available on the polypeptide conformation and three-dimensional structure of any protein pore.

One surface of the membrane-embedded protein must necessarily be hydrophobic to permit its stable interaction with the lipid domain of the bilayer. The other must be hydrophilic to permit the passage of water, ions and small uncharged molecules across the membrane. It is not known whether the protein pores generally span the bilayer, or whether partial penetration sometimes suffices to generate a functional pore. The extent to which pores are lined by protein is also unknown. In the case of the α-toxin hexamer, as well with fully circularized complement and streptolysin-O pores, the channels appear to indeed traverse the interior of the protein cylinders. However, incompletely circularized streptolysin-O pores appear to be lined by an edge of free lipid. It would not be surprising if monomerically inserted proteins such as the hemolysin of *E. coli* do not require circularized protein structures to generate transmembrane channels. Hence, it is possible that part of a pore may be lined by an edge of free lipid which arises through lateral repellment of apolar lipid domains from the inserted, hydrophilic faces of the protein.

Once formed, oligomeric pores are generally very resistent and cannot be destroyed by proteolytic attack at neutral pH. In contrast, monomeric pores created by *E. coli* hemolysin do not exhibit this property. The fate of pores after their formation in membranes of nucleated cells has not been studied in any detail. It is probable that many cells are able to repair a limited number of lesions, either by endocytic uptake or by exocytic shedding of afflicted membrane regions. No studies on the deposition of any pore-forming bacterial cytolysin in human tissue have been performed to date.

Staphylococcal α-Toxin

This toxin is produced by most strains of *S. aureus* and is an important factor of bacterial pathogenicity (14-16). The protein is secreted as a single-chained, nonglycosylated polypeptide of M_r 34,000. The cDNA sequence has been reported (12) but the overall primary structure is unremarkable. The toxin spontaneously binds to lipid mono- (17) and bilayers (18) to produce functional transmembrane pores that have been sized to approximately 1.5 nm diameter (19,20). In the electron microscope, the majority of pores formed at high toxin concentrations (> 20 µg/ml) are seen as circularized rings with central pores of approximately 2 nm diameter. We have proposed that transmembrane leakage is due to embedment of these ring structures in the bilayer, molecular flux occurring through the ultrastructurally visible, central channels (20-22). The rings have been isolated and molecular weight determinations indicate that they represent hexamers of the native toxin (21).

Evidence that α-toxin hexamers generate discrete transmembrane pores is compelling and has been reviewed recently (23,24). We will confine the present discussion to two somewhat controversal issues. The first relates to the question of whether certain cells may possess high-affinity, specific bindings sites for α-toxin which render them highly susceptible to toxin attack. The second is whether membrane damage to such highly susceptible cells is also due to formation of toxin oligomers in the membrane. These two issues are relevant because certain cells (e.g., rabbit erythrocytes) are much more susceptible towards lytic toxin action than others (e.g., human erythrocytes). In an earlier study, the binding of radio-iodinated toxin to rabbit erythrocytes was investigated (25) and the presence of approximately 5000 high-affinity binding sites for α-toxin per cell reported. These sites were absent on human cells. Rabbit erythrocytes lysed by low doses (< 1 μg/ml) of toxin apparently did not display the characteristic hexameric ring structures on their surface. The latter could be detected only at high toxin doses (> 10 μg/ml) such as were also required to lyse human cells. From these findings, it has been surmised that the biological effect of α-toxin ensuing at relatively low (and presumably more physiological) concentrations may be due to interactions with cell surface receptor molecules, and this in turn may initiate a different mode of membrane damage than that observed at high concentrations.

Any binding study conducted with ^{125}I-labeled toxin must, however, be regarded as problematic, unless it can be shown that radiolabeling to sufficiently high specific activity does not cause loss of cytolytic activity. This was not the case in the cited study. For this reason, we elected to use a quantitative immunological approach to investigate toxin binding. A monoclonal antibody generated against α-toxin was found to react only with the monomeric toxin. A sensitive ELISA was constructed with this antibody that permitted detection of only one molecule native toxin per erythrocyte. At the same time, this ELISA allowed simultaneous differentiation and quantitation of monomeric and oligomeric toxin forms on the lysed membranes. Quantitative measurements of toxin binding to rabbit and human erythrocytes were performed over toxin concentrations that ranged from totally noncytolytic doses (< 3 ng/ml) to extremely high doses (100 μg/ml). The following findings were made. First, binding of α-toxin to rabbit erythrocytes was nil at subcytolytic concentrations (approximately 10^{-10}M). In the cytolytic range of 0.05 - 10 μg/ml, net toxin binding remained fairly constant and ineffective, in the order of 5 - 8% of total toxin offered. At very high concentrations (> 50 μg/ml), overall toxin binding increased to reach approximately 40% at 100 μg/ml. These findings confirmed and extended previous data of Arbuthnott (14), Phimister and Freer (26), and ourselves (22), demonstrating that in no range of concentrations did toxin-binding ever exhibit characteristics of a receptor-ligand interaction. Instead, it appears that toxin-binding occurs via nonspecific absorption when a critical concentration threshold is exceeded. The threshold is low in the case of rabbit erythrocytes and high with human cells. The reasons for this are unknown but we believe that nonspecific surface properties of the respective cell targets such as the density and orientation of charged groups are important.

A third finding was that α-toxin binds to cells at 0°C very efficiently, but oligomer formation is greatly retarded. If cells are treated with 0.5 - 1.0 μg/ml toxin at 0°C, the toxin remains in 3S native form and no lysis occurs. When such cells are washed and warmed to 37°C, oligomers rapidly assemble and the cells lyse. These observations demonstrate that membrane damage incurred by low toxin doses involves oligomer formation, and they also show that oligomerization occurs by lateral diffusion and collision of membrane-bound toxin monomers. Further, they have allowed us to estimate that very few oligomers are required to lyse an erythrocyte, compatible with a one-hit-one-hole concept of toxin action.

In conclusion, the latest studies on toxin-membrane interaction clearly argue for a unitarian, pore-forming mode of toxin action and speak against the presence of specific "receptor" molecules on rabbit erythrocytes.

Streptolysin-O

Streptolysin-O (SLO) is the prototype of -SH-activated cytolysins, a group comprising at least 15 different exotoxins that share several properties in common. All are single-chained polypeptides of reported M_r between 40 - 80,000 that can be reversibly inactivated by atmospheric oxygen. Primary binding appears to occur to membrane cholesterol, and cholesterol-free membranes are not susceptible to attack by these toxins. Several of the toxins have been shown to generate circular and semicircular "lesions" of heterogeneous size on target membranes that can be visualized by electron microscopy. Immunological cross-reactivity amongst several toxins in this group has been reported (9-11,27).

The most detailed studies on structure and function have been performed with SLO. This toxin is secreted by Lancefield group A β-hemolytic streptococci as a non-glycosylated molecule of M_r 69,000 and pI = 6.4 (28). That SLO interacts with membrane cholesterol has been inferred from several lines of evidence. Perhaps the most direct has been the demonstration that the toxin binds to soluble cholesterol micelles and becomes inactivated during this process (29). Further, SLO can be precipitated by cholesterol dispersions, and this interaction leads to formation of the typical arc- and ring structures identical to those seen on lysed erythrocyte membranes (30). Functional studies have shown that SLO induces release of markers from liposomes and egress of macromolecules including hemoglobin from erythrocytes (31,32). Since the lesions are very large (> 30 nm) this process is not preceded by osmotic cell swelling, as is found with α-toxin and complement.

The SLO pores have recently been isolated from target erythrocyte membranes and shown to be composed solely of toxin molecules. The pores can be reconstituted into bilayers of pure phosphatidylcholine, where the transmembrane nature of the channels becomes evident (33). The steps leading to pore formation have recently been analyzed with the use of a neutralizing monoclonal antibody (34). This antibody does not inhibit binding of the toxin to cells, but blocks the process of oligomerization. Hence, in the presence of the antibody, SLO remains in monomeric form on erythrocytes and no hemolysis occurs. If cells are then washed and resuspended in buffer without toxin and antibody, slow hemolysis ensues as the monoclonal antibody dissociates and the oligomers form. These results indicate that SLO, like α-toxin, first binds as a monomer to cells, presumably via specific and high-affinity interaction with cholesterol. Subsequent lateral aggregation and collision of membrane-bound toxin then results in oligomerization, which is probably accompanied by unfolding of the molecule and insertion into the bilayer. Basically, the process of pore formation by SLO thus resembles that of α-toxin. Similarly, SLO will also bind to cells at 0°C, but lysis is greatly retarded and requires the presence of high toxin concentrations (9,34).

In summary, both the oligomerizing bacterial cytolysins studied display a similar mode of pore-formation with the basic difference that SLO, because of its interaction with cholesterol, binds much more efficiently to cells than α-toxin. The lateral aggregation mechanism is to be differentiated from a "trigger" mechanism of channel formation such as is found with complement pores. In the latter case, pore formation occurs after a first C9-molecule binds to cell-bound C5b-8, because the C9-molecule becomes a focus for further C9-C9 oligomerization (23). The ultimate outcome of the reaction is similar, i.e., the formation of fully or partially circularized protein pores which display no strict specificity in their interaction with lipids.

The factors that influence the extent of protein oligomerization are unknown. Possibly, quite trivial factors such as membrane surface components (proteins and glycoproteins) act to restrict oligomer formation (toxin collision). An important aspect in this regard is that partially circularized protein channels probably can create functional transmembrane pores because lipid is repelled away from the hydrophilic face to become part of the pore circumference. A tentative proposal for the mode of lipid organization in these regions has been made (33). If confirmed, this type of pore architecture may be present in many other oligomeric and monomeric protein channels.

In closing, attention is drawn to a basic unsolved problem regarding the interaction of sulfhydryl-activated toxins with membrane cholesterol. It remains unclear how such a specific interaction can occur, since the only part of the sterol molecule that is exposed at the membrane surface is its 3β-OH group. Earlier studies have, on the other hand, indicated a requirement for certain membrane-embedded, hydrophobic regions of the sterol molecule for its interaction with SLO (9). How the toxin should recognize such domains is an enigma.

E. coli Hemolysin

This toxin is produced by all hemolytic strains of *E. coli* as a protein of M_r 110,000. The toxin is very unstable, its inactivation possibly being due to changes in its state of aggregation in aqueous solution. The native toxin may be associated with lipopolysaccharide (35) or lipid.

E. coli hemolysin is clearly an important factor of bacterial pathogenicity (36-43). It is produced by *E. coli* in the human host and antibodies are present in sera of healthy adults (44). The hemolysin has been the object of numerous molecular genetical studies. The cDNA sequence is known, as are several aspects of its transport mechanism across the wall of *E. coli* (45-48).

From a functional viewpoint, the hemolysin is interesting because it may represent a new prototype of a membrane-damaging cytolysin that forms monomeric transmembrane pores. This contention is based on several lines of evidence. In erythrocytes, the toxin induces very rapid efflux of K^+ and influx of cations (Ca^{2+}), as well as of small uncharged molecules whose diameters do not exceed 2 nm (49). After membrane solubilization with deoxycholate, the toxin is recovered exclusively in monomer form (49). Furthermore, no ultrastructural lesions can be discerned on erythrocytes lysed with the toxin. In planar lipid bilayers, the toxin creates discrete pores of uniform size (1.5 - 2.0 nm diameter) and the dose-response curves observed during this process exhibit classic one-hit characteristics. This contrasts to the multi-hit dose-response curves in the same system observed with α-toxin. The experiments with planar lipid membranes (Menestrina et al., submitted) have further shown that *E. coli* hemolysin can attack bilayers of pure phosphatidylcholine; hence, this cytolysin falls into the category of receptorless toxins. Finally, this is the first bacterial cytolysin whose action is dependent on membrane potential: channels open when the trans side is negative relative to this cis (application) side, and close when the trans-voltage becomes positive (> 5 mV) relative to the cis-side (Menestrina et al., submitted). In this regard, *E. coli* hemolysin displays highly interesting similarities to pore-forming colicins.

Several basic aspects of toxin action await clarification. These include a quantitative assessment of toxin-binding to cells and the efficiency of pore formation; the influence of membrane composition on toxin action; and possible repair mechanisms by nucleated cells. Many other toxins, particularly those elaborated by gram-negative organisms, may act in a manner similar to *E. coli* hemolysin. Progress in this area should be stimulated by the awareness that transmembrane pore formation need not be synonymous with formation of protein oligomers which are visible by electron microscopy.

Secondary Effects Elicited by Pore-Forming Cytolysins

Stimulus-Response Coupling Effects. Since toxin pores and nonselective or cation-selective, they permit rapid transmembrane flux of Ca^{2+} and other small molecules into the cells. Incoming calcium will trigger diverse intracellular events including stimulation of arachidonate metabolism (50). In the model of the isolated and perfused rabbit lung, addition of α-toxin to the re-circulating buffer involved development of pulmonary hypertension that was correlated to the synthesis of prostanoids (51). The lung model bears potential significance because the observed pathophysiological process may contribute to the pathogenesis of acute respiratory distress and related syndroms during toxinemia. It is noteworthy that *E. coli* hemolysin, in addition to eliciting marked pulmonary hypertension in the isolated lung, also induces a rise in vascular permeability and causes pulmonary edema (Seeger et al., unpublished data). The mechanisms involved in the latter phenomenon are under study. Further search will undoubtedly reveal other secondary effects induced by molecular flux across toxin pores in nucleated cells.

Immunological Self-Attack at the Membrane Level. All pore-formers studied to date are excellent immunogens in the human host. In their cell-bound form, they will therefore bind host antibody and in some cases this will induce complement activation. The first demonstration of such immunological self-attack at the membrane level was made with SLO, which induces complement activation in an antibody-dependent manner in vitro (52). This leads to rapid complement consumption, generation of C3a and C5a anaphylatoxins, binding of large amounts of C3b which will act as an opsonin, and also to generation of C5b-9 complement lesions in autologous cell membranes.

Such processes, if occurring in vivo, would support in situ inflammation as well as opsonization and phagocytosis of host target cells. In broader perspective, immunological self-attack induced by bacterial molecules that enter host membranes may be compared to T-cell attack on virus-infected cells that bear alien antigens on their surface. An analogous complement-dependent self-attack mechanism had recently been described after spontaneous insertion of bacterial lipoteichoic acid into cell membranes (53).

Conclusions

The concept of membrane damage by formation of transmembrane pores through insertion of alien proteins into target bilayers is today well established in many biological systems. The list of pore formers will certainly grow as general awareness of this mechanism increases. Interesting aspects of protein and membrane biochemistry, cell physiology and immunology are brought together by the study of these proteins. Seen in broader perspective, the phenomenon of a hydrophilic-amphiphilic transition of proteins appears to be essential in a spectrum of other important biological events. These include the pH-dependent escape of toxin fragments (54,55) and virus genomes from endosomes (56, 57), and the passage of mitochondrial and chloroplast proteins through respective membranes (58,59). From a biological viewpoint, it is clear that membrane damage by pore-forming proteins is a very widespread phenomenon that plays a central role both in the context of microbial pathogenesis and as an effector mechanism of mammalian defence systems.

Acknowledgment. Work performed in the author's laboratories was supported by the Deutsche Forschungsgemeinschaft (Bh 2/2 and Bh 2/1-7) and the Verband der Chemischen Industrie.

References

1. Bhakdi S, Tranum-Jensen J (1978) Molecular nature of the complement lesion. Proc Natl Acad Sci USA 75:5655-5659
2. Mayer MM, Michaels DW, Ramm LE, Whitlow MB, Willoughby JB, Shin ML (1981) Membrane damage by complement. CRC Crit Rev Immunol 7:133-165
3. Bhakdi S, Tranum-Jensen J (1983) Membrane damage by complement. Biochim Biophys Acta 737:343-372
4. Henkart P (1985) Mechanism of lymphocyte-mediated cytotociscity. Annu Rev Immunol 3:31-58
5. Faulstich H, Bühring HJ, Seitz J (1983) Physical properties and function of phallolysin. Biochemistry 22:4574-4580
6. Lynch EC, Rosenberg IM, Gitler C (1982) An ion-channel forming protein produced by *Entamoeba histolytica*. EMBO J 1:801-804
7. Young JDE, Young TM, Lu LP, Unkeless JC, Cohn ZA (1982) Characteriaztion of a membrane pore-forming from *Entamoeba histolytica*. J Exp Med 156:1677-1690
8. Michaels DW (1979) Membrane damage by a toxin from the sea anemone stoichactis helianthus. I. Formation of transmembrane channels in lipid bilayers. Biochim Biophys Acta 555:67-78
9. Alouf JE (1980) Streptococcal toxins (streptolysin-O, streptolysin-S, erythrogenic toxin). Pharmacol Ther 11:661-717
10. Bernheimer AW (1974) Interactions between membranes and cytolytic bacterial toxins. Biochim Biophys Acta 344:27-50
11. Smyth CJ, Duncan JL (1978) Thiol-activated (oxygen-labile) cytolysins. In: Jeljaszewicz J, Wadström T (eds) Bacterial toxins and cell membranes. Academic Press, London New York, pp 129-183
12. Gray GS, Kehoe M (1984) Primary sequence of the α-toxin gene from *Staphylococcus aureus* wood 46. Infect Immun 46:615-618
13. Felmlee T, Pellet S, Welch R (1985) Nucleotide sequence of an *Escherichia coli* chromosomal hemolysin. J Bacteriol 163:94-105
14. McCartney C, Arbuthnott JP (1978) Mode of action of membrane-damaging toxins produced by staphylococci. In: Jeljaszewicz J, Wadström T (eds) Bacterial toxins and cell membranes. Academic Press, London New York, pp 89-127
15. Johnsson P, Lindberg M, Haraldsson I, Wadström T (1985) Virulence of *Staphylococcus aureus* in a mouse masititis model: studies of alpha haemolysin. Coagulase and protein A as possible virolence determinants with protoplast fusion and gene cloning. Infect Immun 49:765-769

16. O'Reilly M, Azavedo JCS, Kennedy S, Foster TJ (1986) Inactivation of the alpha-haemolysin gene of *Staphylococcus aureus* 8325-4 by site-directed mutagenesis and studies on the expression of its haemolysins. Microbial Pathogen 1:125-138
17. Buckelew AR, Colaccico G (1971) Lipid monolayers. Interaction with staphylococcal α-toxin. Biochim Biophys Acta 233:7-16
18. Freer JH, Arbuthnott JP, Bernheimer AW (1968) Interaction of staphylococcal α-toxin with artificial and natural membranes. J Bacteriol 95:1153-1168
19. Menestrina G (1986) Ionic channels formed by *Staphylococcus aures* α-toxin: voltage-dependent inhibition by divalent and trivalent cations. J Membr Biol 90:177-190
20. Füssle R, Bhakdi S, Sziegoleit A, Tranum-Jensen J, Kranz T, Wellensiek HJ (1981) On the mechanism of membrane damage by *S. aureus* α-toxin. J Cell Biol 91:83-94
21. Bhakdi S, Füssle R, Tranum-Jensen J (1981) Staphylococcal α-toxin: oligomerisation of hydrophilic monomers to form amphiphilic hexamers induced through contact with deoxycholate detergent micelles. Proc Natl Acad Sci USA 78:5474-5479
22. Bhakdi S, Muhly M, Füssle R (1984) Correlation between toxin binding and hemolytic activity in membrane damage by staphylococcal α-toxin. Infect Immun 46:318-323
23. Bhakdi S, Tranum-Jensen J (1987) Damage to mammalian cells by proteins that form transmembrane pores. Rev Physiol Biochem Pharmacol
24. Bhakdi S, Tranum-Jensen J (1987) Damage to cell membranes by pore-forming bacterial cytolysins. Prog Allergy (in press)
25. Cassidy P, Six HR, Harsham S (1976) Studies on the binding of staphylococcal ^{125}I-labelled α-toxin to rabbit erythrocytes. Biochemistry 15:2348-2355
26. Phimister GM, Freer JH (1984) Binding of ^{125}I-alpha toxin of *Staphylococcus aureus* to erythrocytes. J Med Microbiol 18:197-204
27. Cowell JL, Bernheimer AW (1977) Antigenic relationships among thiol-activated cytolysins. Infect Immun 16:397-399
28. Bhakdi S, Roth M, Sziegoleit A, Tranum-Jensen J (1984) Isolation and identification of two hemolytic forms of streptolysin-O. Infect Immun 46:394-400
29. Johnson MK, Geoffroy C, Alouf JE (1980) The binding of cholesterol by sulfhydryl-activated cytolysins. Infect Immun 27:97-101
30. Duncan JL, Schlegel R (1975) Effect of streptolysin O an erythrocyte membranes, liposomes, and lipid dispersions. A protein-cholesterol interaction. J Cell Biol 67:160-173
31. Duncan JL (1974) Characteristics of streptolysin O hemolysis: kinetics of hemoglobin and ^{86}rubidium release. Infect Immun 9:1022-1027
32. Buckingham L, Duncan JL (1983) Approximate dimensions of membrane lesions produced by streptolysin-S and streptolysin-O. Biochim Biophys Acta 729:115-122
33. Bhakdi S, Tranum-Jensen J, Sziegoleit A (1985) Mechanism of membrane damage by streptolysin-O. Infect Immun 47:52-60
34. Hugo F, Reichwein J, Arvand M, Krämer S, Bhakdi S (1986) Mode of transmembrane pore formation by streptolysin-O analysed with a monoclonal antibody. Infect Immun 54:641-645
35. Bohach GA, Snyder IS (1985) Chemical and immunological analysis of the complex structure of *E. coli* hemolysin. J Bacteriol 164:1071-1080
36. Cavalieri SJ, Bohach GA, Snyder IS (1984) *Escherichia coli* α-hemolysin: characteristics and probable role in pathogenicity. Microbial Rev 48:326-343
37. Welch RA, Patchen-Dellinger E, Minshew B, Falkow S (1981) Hemolysin contributes to virulence of extraintestinal *E. coli* infections. Nature (London) 295:665-667
38. van den Bosch JF, Postum P, Koopman PAR, de Graaff J, MacLaren CM, van Brenk DG, Guinee PAM (1982) Virulence of urinary and faecal *Escherichia coli* in relation to serotype, haemolysis and haemagglutination. J Hyg 88:567-577
39. Welch RA, Hall R, Falkow S (1983) Molecular cloning and physical characterisation of a chromosomal haemolysin from *Escherichia coli*. Infect Immun 42:178-186
40. Waalwijk C, MacLaren DM, De Graaff J (1983) In vivo function of hemolysin in the nephropathogenicity of *Escherichia coli*. Infect Immun 42:245-249
41. Hughes C, Hacker J, Robert A, Goebel W (1983) Hemolysin production as a virulence marker in symptomatic and asymptomatic urinary infections caused by *Escherichia coli*. Infect Immun 39:546-551
42. Hacker J, Hughes C, Hof H, Goebel W (1983) Cloned hemolysin genes from *Escherichia coli* that cause urinary tract infection determine different levels of toxicity in mice. Infect Immun 42:57-63

43. Welch RA, FAlkow S (1984) Characterisation of *Escherichia coli* hemolysins conferring quantitative differences in virulence. Infect Immun 43:159-160
44. Hugo F, Arvand M, Reichwein J, Mackman N, Holland IB, Bhakdi S (1987) Identification of hemolysin produced by clinical isolates of *E. coli* with monoclonal antibodies. J Clin Microbial 25:26-30
45. Goebel W, Hedgepeth J (1982) Cloning and functional characterization of the plasmid-encoded hemolysin determinant of *Escherichia coli*. J Bacteriol 151: 1290-1298
46. Mackman N, Holland IB (1984) Functional characterization of a cloned haemolysin determinant from *E. coli* of human origin, encoding information for the secretion of a 107K polypeptide. Mol Gen Genet 196:123-134
47. Felmlee T, Pellett S, Lee EY, Welch R (1985) *Escherichia coli* haemolysin is released extracellularly without cleavage of a signal peptide. J Bacteriol 163: 88-93
48. Mackman N, Nicaud JM, Gray L, Holland IB (1986) Secretion of hemolysin by *E. coli*. Curr Top Microbiol Immunol 125:159-181
49. Bhakdi S, Mackman N, Nicaud JM, Holland IB (1986) *E. coli* hemolysin may damage target cell membranes by generating transmembrane pores. Infect Immun 52:63-69
50. Suttorp N, Seeger W, Dewein E, Bhakdi S, Roka L (1985) Staphylococcal α-toxin stimulates synthesis of prostacyclin by cultured endothelial cells from pig pulmonary arteries. Am J Physiol 248:C127-135
51. Seeger W, Bauer M, Bhakdi S (1984) Staphylococcal α-toxin elicits hypertension in isolated rabbit lungs due to stimulation of the arachidonic acid cascade. J Clin Invest 74:849-858
52. Bhakdi S, Tranum-Jensen J (1985) Complement activation and attack on autologous cells induced by streptolysin-O. Infect Immun 48:713-719
53. Hummell DS, Winkelstein JA (1986) Bacterial lipoteichoic acid sensitizes host cells for destruction by autologous complement. J Clin Invest 77:1533-1538
54. Sandvig K, Olsnes S (1981) Rapid entry of nicked diphtheria toxin into cells at low pH. J Biol Chem 256:9068-9076
55. Kagan BL, Finkelstein A, Colombini M (1981) Diphtheria fragment forms large pores in phospholipid bilayer membranes. Proc Natl Acad Sci USA 78:4950-4954
56. Helenius A, Kartenbeck J, Simons K, Fries E (1980) On the entry of Semliki Forest Virus into BHK-21 cells. J Cell Biol 84:404-420
57. Skehel JJ, Bayley PM, Brown EB, Martin SR, Waterfield MD, White JM, Wilson IA, Wiley DC (1982) Changes in the conformation on influenza virus hemagglutinin at the pH-optimum of virus-mediated membrane fusion. Proc Natl Acad Sci USA 79:968-972
58. Sabatini DD, Kreibich G, Morimoto T, Adesnik M (1982) Mechanisms for the incorporation of proteins in mebranes and organelles. J Cell Biol 92:1-22
59. Schmidt GW, Mishkind ML (1986) The transport of proteins into chloroplasts. Annu Rev Biochem 55:879-912

Bacterial Cytolysins – Extracellular Proteins and Virulence Factors

S. Kathariou, J. Hacker, H. Hof, I. Then, W. Wagner, M. Kuhn, and W. Goebel[1]

Introduction

Synthesis of hemolysis has been reported throughout a broad spectrum of gram-negative and gram-positive bacterial species. The in vitro action of hemolysins ultimately causes physical damage to a variety of eukaryotic cells. Bacterial products possessing such a membrane-damaging action have been termed cytolysins and for many of the hemolysins, a general cytolytic action on mammalian cells has been demonstrated.

The introduction of genetic techniques, especially recombinant DNA techniques, has accelerated detailed studies of hemolysin synthesis and transport in a number of bacteria. The isolation of the genes and their gene products, in conjunction with animal models, has helped to elucidate the contribution of hemolysin to the virulence of hemolytic strains. The most extensive studies on hemolysins have been performed in *Escherichia coli* (1,2), but recently progress has been also made in the genetic analysis of other hemolysins from both gram-positive and gram-negative bacteria. Although more hemolytic extracellular proteins are described and characterized in gram-positive than in gram-negative bacteria, less work has been reported on the genetics of these hemolysins. This reflects the difficulties encountered in cloning hemolysin genes from gram-positive bacteria. Cloning of such genes in *E. coli* is often accompanied by problems of limited transport, possibly incorrect processing of putative precursors and, at times, of imperfect gene expression. Here we describe molecular studies on the hemolysins (cytolysins) of two bacterial systems, namely *Escherichia coli* and *Listeria monocytogenes*, a gram-positive intracellular pathogen. It is further shown that both cytolysins represent important virulence factors.

Results and Discussion

I. The Genetic Determinant for hemolysin from *Escherichia coli*

The genetics of alpha hemolysin synthesis in *E. coli* was intensively studied during the past years (for review see 1,2). Hemolysin synthesis is mainly observed in *E. coli* strains causing extraintestinal infections and the incidence of hemolytic strains is particularly high in *E. coli* strains causing pyelonephritis. These strains carry the *hly* determinant predominantly on the chromosome, whereas many hemolytic *E. coli* isolates from animal source harbor plasmid-encoded *hly* determinants. The chromocomal and the extrachromosomal *hly* determinants are highly homologous and consist of four structural genes (Fig. 1), designated as *hly*C, A, B, and D (3). These four *hly* genes are controlled by at least two promoters. One promoter (p*hly*$_a$) is located upstream of

[1]Institut für Genetik und Mikrobiologie der Universität Würzburg, D-8700 Würzburg, FRG

38. Colloquium Mosbach 1987
Molecular Basis of Viral and Microbial Pathogenesis

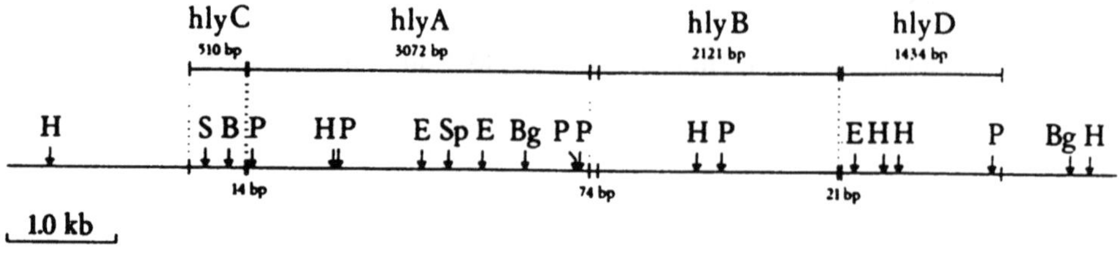

Fig. 1. Hemolysin determinant of *E. coli*, consisting of the four genes, *hly*C, *hly*A, *hly*B and *hly*D. *Numbers* indicate the nucleotides of the structural genes and the intergenic regions

*hly*C and at least another one (p*hly*$_b$) in front of *hly*D. Transcription of *hly*C, *hly*A and possibly *hly*B starts from p*hly*$_a$. A sequence of more than 400 bp upstream of *hly*C has been shown to be necessary for full expression of a chromosomal *hly* determinant (4). This regulatory region of most extrachromosomal *hly* determinants differs considerably in its nucleotide composition from the corresponding chromosomal sequences (5,6), and seems to be interrupted in most Hly plasmids by an IS2 insertion at position -470 bp upstream of *hly*C (position 1 being the first nucleotide of the *hly*C gene) since an additional sequence of maximally 400 bp upstream of IS2 which acts in cis is necessary for full expression of the plasmid-encoded *hly* genes. The precise function of this sequence (*hly*R) is unknown but it enhances synthesis of active hemolysin 25- to 50-fold and secretion of hemolysin more than 100-fold (compared to the *hly*R$^-$ situation (Fig. 1). The plasmid-encoded *hly* determinant is flanked in most Hly plasmids by insertion-type elements (2) whereas chromosomal *hly* determinants may be located on larger chromosomal inserts (70 kb to 100 kb) which carry short directly repeated sequences at their ends (7). Homologous recombination between the two direct repeats leads to deletions which remove these inserts precisely. The nephropathogenic *E. coli* 06 strain 536 carries two such inserts of 75 kb and 100 kb in size. Each of them contains an active *hly* determinant (*hly*I and *hly*II). Spontaneous deletion mutants were obtained from this strain which suffered deletions of either one (*hly*I or *hly*II) or both inserts (8,9).

Synthesis and Secretion of E. coli Hemolysin. Two *hly* determinants, one of chromosomal origin and another of plasmid origin, were sequenced (6, 13) and the amino acid sequences of the four *Hly* proteins were deduced from observed ORF's. The secreted hemolysin consists of the unprocessed *Hly*A product, a protein of 110,000 Da. This protein is found in the supernatant only when the two *hly* genes B and D are expressed. In the absence of *hly*C, the HlyA protein is likewise secreted by HlyB and HlyD but this HlyA is hemolytically inactive. This suggests that HlyC protein activates HlyA by a yet unknown mechanism to the hemolytically active form.

Recent data show that active hemolysin is extremely sensitive to phospholipase C (Fig. 2). This phospholipase rapidly inactivates extra- and intracellular hemolysin without affecting the size of the HlyA protein on SDS-PAGE (Fig. 2). Eryhtrocytes treated with this lipase are not lyzed, nor do they become resistant to hemolysin. This is evidence that association of HlyA with phospholipid(s) may be essential for activity. Whether HlyC is involved in this step remains to be clarified.

The proteins, HlyB and HlyD, both of which seem to represent transmembrane proteins (6, Goebel, Hess and Vogel, unpublished) form a specific transport system for HlyA. HlyA does not possess a conventional

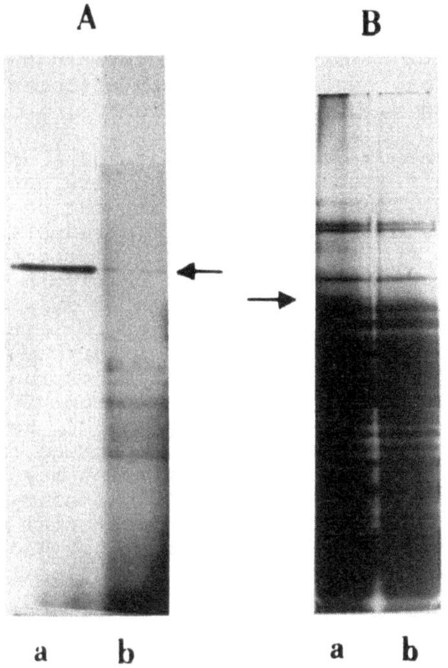

Fig. 2A,B. Extracellular (A) and intracellular HlyA protein (110 kDa) synthesized in the presence *(a)* and absence *(b)* of *hly*R. The arrows indicate the position of HlyA protein on the SDS-polyacrylamide gels. Equal amounts of supernatants in A (equivalent to 3 ml) and cellular extracts in B were applied to the gels

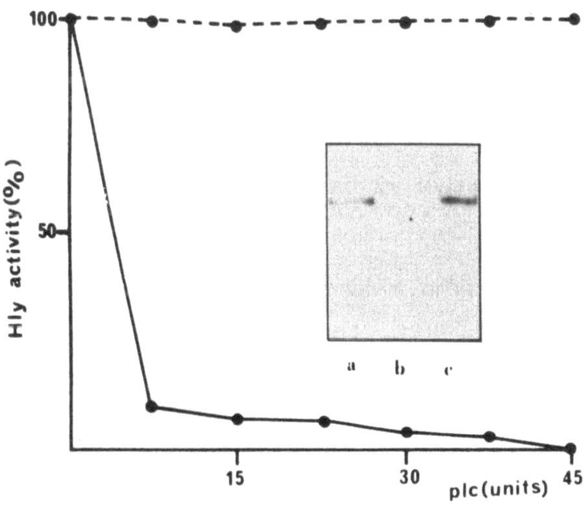

Fig. 3. Phospholipase sensitivity of extracellular hemolysin. A hemolysin preparation (containing about 2 µg protein HlyA) was treated for 3 min with the indicated concentrations of phospholipase C (plc) and hemolytic activity was measured as described (3). The *solid line* shows the decrease of hemolytic activity after treatment with phospholipase C, whereas the *dashed line* shows the activity after treatment of hemolysin under the same conditions without phospholipase C. The insert *(a–c)* represents HlyA protein after chromatography on PAGE of untreated hemolysin *(a)*, hemolysin treated with 45 units of phospholipase C *(c)* and hemolysin treated with 1 unit of proteinase K for 5 min *(b)*

N-terminal signal peptide. Yet there is recent evidence that the N-terminal sequence of HlyA, which resembles a transit sequence required for import of nuclear proteins into organelles (14), is membrane-active, since proceins derived from fusion between *hly*A and *pho*A, the latter being devoid of the transport signal (by TnphoA; (15)), transport at least the PhoA part across the cytoplasmic membrane when less than 100 amino acids from the N-terminal end of HlyA are fused to PhoA (Fig. 3). However, transport of these HlyA • PhoA fusion proteins across the outer membrane does not occur even in the presence of hlyB and hlyD. A possible recognition signal on HlyA for the HlyB/HlyD transport system seems to be located within the last 175 amino acids of the C-terminal part of HlyA ((16), W. Wagner, A. Ludwig and W. Gobel, unpublished). PhoA fusion proteins containing in addition to the above described N-terminal sequence of HlyA, this C-terminal sequence are very efficiently transported by HlyB/HlyD (J. Hess and W. Goebel, unpublished).

E. coli Hemolysin as Virulence Factor. Spontaneous deletion mutants of *E. coli* 536 which have lost both *hly* determinants (see above) are nonvirulent when tested in several amino models (10). This type of mutant can be used as suitable host for reconstitution of virulence by reintroduction of cloned virulence genes. Hemolysin determinants from various origins were introduced in such mutants. The results indicate that hemolysin alone exhibits toxic effects after injection of the manipulated strains into the mouse peritoneum or in chicken embryos. The level of toxicity depends on the used *hly* determinant and is in general higher for *hly* determinants which are of chromosomal origin than for those of extra-chromosomal origin. The molecular cause for the difference in toxicity lies in the amount of extracellular hemolysin expressed by the *hly* determinant (which is mainly a function of sequences located upstream of *hly*C) and in the quality of the secreted hemolysin protein which may differ in amino acid composition between hemolysin from various hemolytic *E. coli* strains (2). In addition, the serotype of the bacterial host seems to influence the toxic potential of *E. coli* hemolysins (Hacker et al., unpublished). With these genetically manipulated *E. coli* strains it has been further shown that sublytic doses of hemolysin induce in granulocytes and Mast cells the synthesis of leucotrienes and the release of these inflammatory substances and of performed histamine (11, 2).

II. Hemolysin (Listerolysin) from *Listeria monocytogenes* as Virulence Factors

Listeria monocytogenes is a gram-positive, rod-shaped bacterium which can cause severe infections in man and animals. *L. monocytogenes* can invade macrophages and monocytes, as well as nonprofessional phagocytic cells (see below) and is able to survive and to multiply within these cells. It thus belongs to the group of intracellular bacteria. Most strains of *L. monocytogenes* are producers of a beta-hemolysin, termed listeriolysin. This hemolysin (cytolysin) has been characterized as an extracellular protein that belongs to the group of SH-activated, cholesterol-inhibited cytolysins and crossreacts immunologically with streptolysin O (SLO) (17). There is circumstantial evidence that more than one type of hemolysin may exist in strains of *Listeria* (18), (Leimeister and Chakraborty, unpublished).

To demonstrate the role of listeriolysin as a virulence factor, the genetic determinant essential for this protein in *L. monocytogenes* (Mackaness strain serotype 1/2a) has been inactivated by insertion of the tetracycline resistance transposon Tn916 into *L. monocytogenes* DNA (19). The transposon was transferred by means of conjugation of a streptomycin-resistant *L. monocytogenes* recipient strain with *Streptococcus faecalis* CG110 on membrane filters (20). Among the tetracycline-resistant transconjugants, several mutants were detected which has lost hemolytic activity. Hybridization of chromosomal DNA's from these mutants with Tn916 DNA as a probe indicates that in the majority of them the transposon is present on a single chromosomal *Eco*RI fragment (Tn916 does not contain an *Eco*RI site) and two *Hind*III fragments (Tn916 contains one *Hind*III site). This suggests that a single transposition event can lead to the loss of the hemolytic phenotype. Comparative analysis of the extracellular proteins from supernatants of wild-type *L. monocytogenes* 1/2a and of several Tn916 Hly$^-$ mutants indicates the complete loss of

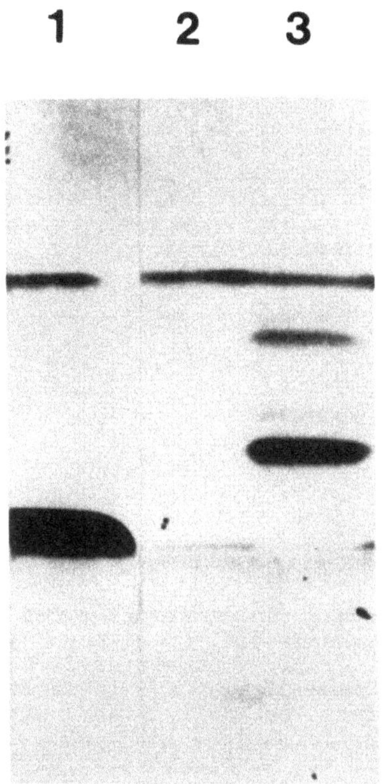

Fig. 4. Immunoblot showing the crossreaction of the extracellular proteins of *L. monocytogenes* with anti-streptolysin O. Lanes: *1* Hly$^-$ mutant M13, producing a presumptive truncated hemolysin; *2* Hly$^-$ mutant M3; *3* Hly$^+$ Tn916 carrying strain WT. Note the strongly reacting band of 49 kDa observed in lane *1*. The lower, lightly reacting bands observed for this mutant were probably due to proteolytic degradation of the 49 kDa peptide. The lowest band, observed with equal intensity in the case of Hly$^-$ and Hly$^+$ strains, was probably due to unspecific binding of the antibodies

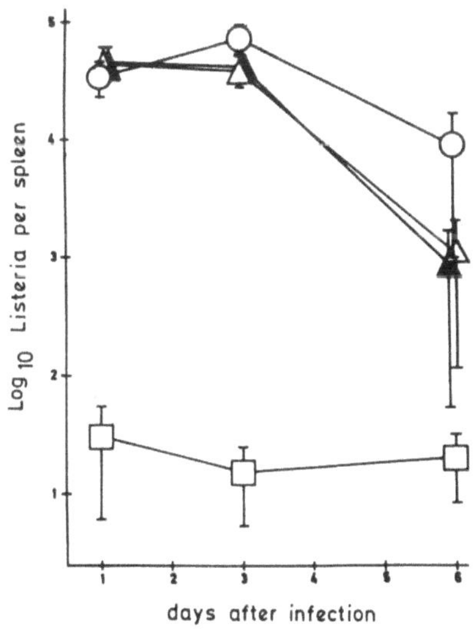

Fig. 5. Course of infection by wild-type and Hly$^-$ *L. monocytogenes*. Mice were inoculated intravenously with 1.4×10^3, 2.7×10^3, and 3.2×10^3 bacteria of the Mackaness (O), WT (Δ), and M3 (□) strains, respectively. The *numbers* are bacterial counts per spleen 1, 3, and 6 days after infection. Each number is the mean from four or five animals. In the case of infections by WT, bacteria were also plated on medium with tetracycline as indicated (▲)

a single major protein of 58 kDa. Since this protein crossreacts specifically with SLO antiserum and binds specificially to erythrocyte membranes, we presume it to be listeriolysin (Fig. 4). In another type of Hly$^-$ mutant, smaller proteins appear instead of the 58 kDa protein

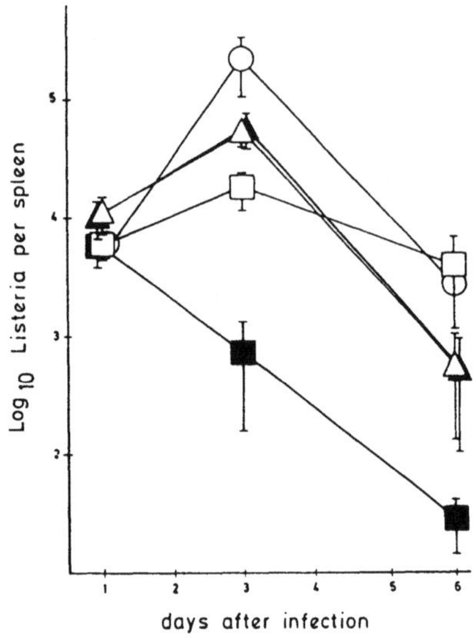

Fig. 6. Recovery of virulence by hemolytic revertants of the Hly⁻ mutant. Samples of 2.7×10^3 2.7×10^3, and 1.4×10^6 bacteria of the Mackaness (O), WT (Δ), and M3 (□) strains, respectively, were injected intravenously. Bacterial counts per spleen were determined 1, 3, and 6 days after infection. In the case of infections with WT and M3, bacteria were also plated on media with tetracycline (▲ and ■, respectively). The *numbers* are means from five animals

which cross react with SLO antiserum and probably represent truncated listeriolysin proteins (Fig. 5). The size of the *Hind*III fragments hybridizing with the Tn916 DNA probe in the latter Hly^- mutants differs only slightly. This suggests that Tn916 has been in each case inserted in chromosomal sites which are in close vicinity (maximally 1 kb apart). This is consistent with the idea that at least in this mutant type Tn916 has been inserted in the structural gene for listeriolysin. Hly^+ revertants from both types of Tn916-bearing Hly^- mutants were obtained which has lost the entire Tn916 probably by a precise excision. These revertants synthesize again normal levels of the 58 kDa protein which reacts with SLO antiserum as expected for listeriolysin.

Following intravenous injection of the wild-type *L. monocytogenes* (Mackaness strain) into mice, the course of infection previously desribed by Mackaness (21) for experimental infections of mice with virulent *L. monocytogenes* was observed. Following an initial reduction in numbers, the number of bacteria isolated from the spleens of the animals started increasing, reached a peak between the third and fourth day after infection and then declined (Fig. 5). The number of bacteria isolated from the spleens of the animals three days after infection was 50 times higher than that present in the inoculum. Intravenous infection of a Tet^+, Hly^+ transposon-bearing WT strain (wild type with respect to hemoylsin production) resulted in a similar pattern of bacterial multiplication within the animals, although the extent of multiplication was somewhat lower than that observed in the case of the parental Mackaness strain and a 13-fold increase was observed. Intravenous injection of the Tet^+, Hly^- mutant bacteria, however, was not followed by multiplication of the bacteria in the animals; instead, the numbers of bacteria isolated from the spleen declined sharply after their introduction and as soon as 1 day later the bacteria were virutally eliminated from the animals (Fig. 5). The few bacteria that could be isolated from the spleens were resistant to tetracycline and nonhemolytic. Similar results were observed in the experimental infections of mice by other Hly^- mutants and randomly selected Hly^+ transconjugants produced from different crosses.

The infection dose of Hly^- bacteria that was used in these experiments (3.1×10^3 bacteria) is unlikely to contain any Hly^+ revertants, since

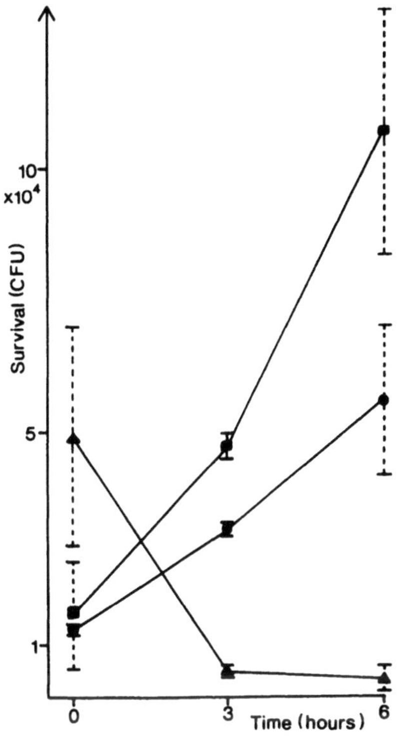

Fig. 7. Persistance of *L. monocytogenes* wild-type strain and Hly$^-$ mutant within peritoneal macrophages. 5×10^5 macrophages were infected with 5×10^6 bacteria and uptake and survival of the bacteria were determined as described in Material and Methods. Each *point* in the figure represents an average of two experimental values. The deviations of the two values are indicated by the *dashed lines*. ■ *L. monocytogenes* (Mackaness wild-type strain Hly$^+$, ● *L. monocytogenes* (Mackaness) Hly$^+$ Tc^r (due to a Tn916 insertion - ref. 13), ▲ *L. monocytogenes* (Mackaness) Tc^r, Hly$^-$ (due to a Tn916 insertion in the structural *hly* gene (13))

the average reversion rate in vitro is 10^{-4}. We expected that if such revertants were generated they would regain their ability to multiply within the animals. Moreover, on the basis of the in vitro data we expected that most of these revertants would lose resistance to tetracycline. To confirm these expectations we repeated the experimental infections of mice with the three strains (the Mackaness wild type strain and the Tn916-bearing derivatives; Tet^r, Hly$^+$ strain WT and Tet^r, Hly$^-$ strain M3) using standard sublethal doses of the hemolytic strains but three orders of magnitude more bacteria of the Hly$^-$ mutant (1.3×10^6) (Fig. 6). The multiplication pattern of the hemolytic strains was similar to that observed earlier. The bacteria isolated from the spleens of the animals infected with strain WT were all Hly$^+$, and similar numbers were isolated on media with and without tetracycline, suggesting that in vivo the bacteria maintained their resistance to tetracycline and that loss of the transposon had not occurred. Most of the bacteria isolated from the animals which had been infected with the Hly$^-$ strain M3, however, were hemolytic and sensitive to tetracycline. Such bacteria represented 40, 86, and 100% of all those which were isolated 1, 3, and 6 days after infection, respectively. The Hly$^+$, Tet^6 bacteria were apparently revertants which had lost the transposon and regained the hemolytic phenotype, along with the ability to multiply within the animals. Only a small fraction of the bacteria isolated from these animals were still Hly$^-$, suggesting that even when injected in high dosages the Hly$^-$ bacteria were unable to multiply within the animals and were rapidly eliminated. These data strongly suggest that the hemolysin (listeriolysin) of *L. monocytogenes* is an essential virulence factor.

In order to examine the positive involvement of the hemolysin in the intracellular survival of the bacteria we were studying the interactions

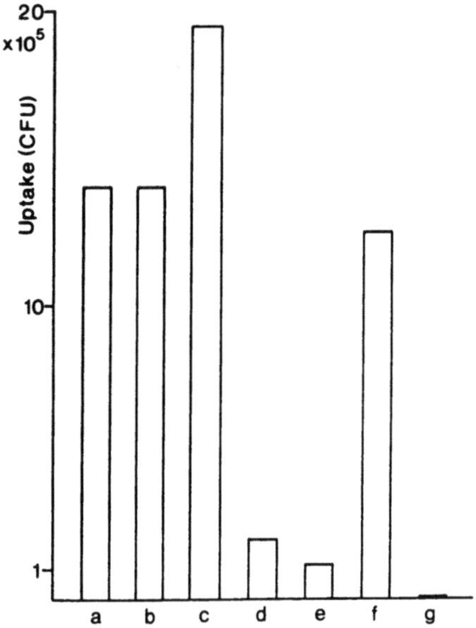

Fig. 8. Invasion of different *Listeria* strains in 3T6 fibroblast cells. Invasion assays were performed essentially as described (11). Each column represents the mean value of two experiments. *a L. monocytogenes* (Mackaness); *b L. monocytogenes* (EGD); *c L. monocytogenes* SV4b; *d L. monocytogenes* SV3a; *e L. innocua* SV6a; *f L. monocytogenes* (Mackaness) Tc^R, Hly^-, *Bacillus subtilis* BR151

between the Hly^+ and Hly^- bacteria and phagocytic cells in vitro. Mouse peritoneal macrophages were prepared as described (22) and unopsonized *Listeria* cells were added in a ratio of ten *Listeria* cells per macrophage (Fig. 7). Following an initial 40 min incubation the bacteria that remained outside the macrophages were killed by gentamicin and the number of viable bacteria within the macrophages was determined at time 0 and at 3 h and 6 h thereafter. Within the first 3 h the numbers of both Hly^+ and Hly^- bacteria initially decreased to 50% of those present originally (time 0). After 3 h inside the macrophages, however, the Hly bacteria started increasing, reaching at 6 h twice the numbers present originally. In contrast to this, the Hly^- bacteria failed to multiply inside the macrophages and their numbers declined steadily, so that at 6 h only 20% of the bacteria present at time 0 remained. These data suggest that the ability of the bacteria to survive and multiply within the macrophages is severely impaired in the Hly^- mutants. Hly^+ and Hly^- *Listeria monocytogenes* are identical in terms of their ability to be internalized by nonprofessional phagocytes (Fig. 8). Following centrifugation with 3T1 cell monolayers Hly^+ and Hly^- bacteria were able to be internalized by the cells with similar frequencies. Internalization is probably due to phagocytosis, since it is strongly inhibited by cytochalasin B, an inhibitor of microfilament function. In contrast to the Hly^- mutants of *L. monocytogenes* two other *Listeria* strains, the avirulent nonhemolytic *L. innocua* (SV6a) and the slightly virulent, hemolytic *L. monocytogenes* (SV3a) do not enter these host cells to any appreciable extent similar to *B. subtilis* which was used as a noninvasive control bacterium. These data clearly show that fully virulent *Listeria* must possess an active mechanism for invasion of mammalian cells. Listeriolysin is clearly not involved in the invasion step but there is circumstantial evidence that another extracellular protein with a relative molecular mass of 60,000 Da may be essential for invasion. This protein is still synthesized by all Tn916-induced Hly^- mutants of *L. monocytogenes* but absent in all nonvirulent *Listeria* strains tested and absent or considerably reduced in amount in certain rough mutants of *L. monocytogenes* (Fig. 9).

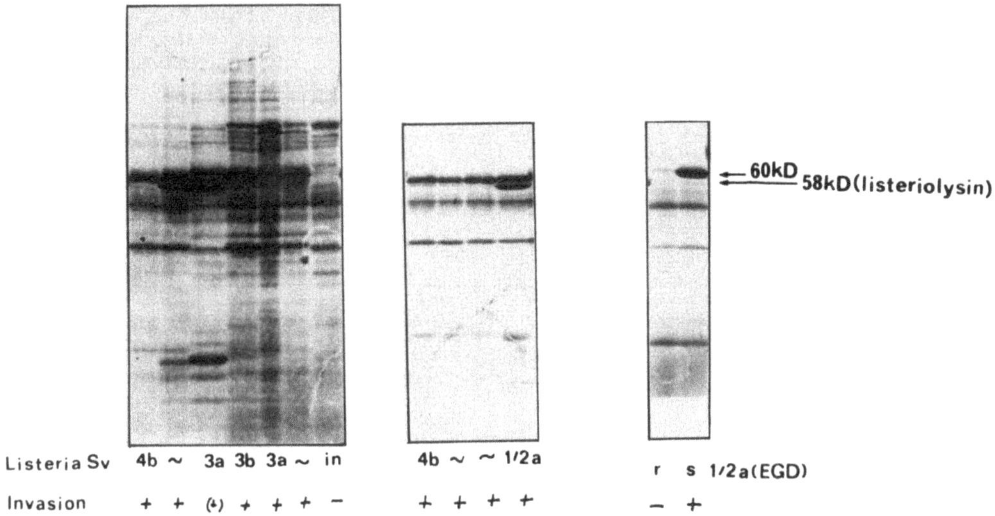

Fig. 9. Correlation between the capability of invading mammalian cells and the possesion of a 60 kDa extracellular protein by *Listeria* strains. Invasion was tested with 3T6 cells as before using virulent and avirulent *Listeria* strains of various serovars *(Sv)*; *r* designates a spontaneous rough mutant of the smooth *(s) L. monocytogenes* EGD strain. Note the reduced amount (and the possibly altered size) of the 60 kDa protein in the *3a* strain in lane *3* and the virtual absence of this protein in *Listeria innocua (in)* and the rough mutant *(r)* of the *L. monocytogenes* EGD strain. The amount of listeriolysin (crossreacting with anti-SLO antiserum) at 58 kDa varies significantly among these strains

Conclusions

The two virulent bacteria described in this study, *Escherichia coli* and *Listeria monocytogenes*, have in common the potential of producing extracellular hemolytic toxins. The complexity of the genetic determinants encoding these hemolysins (cytolysins) and the physical and biochemical properties of the corresponding exoproteins are different. Nevertheless, each represents a potent virulence factor, since its loss reduces drastically the virulence of the respective bacterial strain. Although both hemolysins are able to quickly lyze erythrocytes in vitro (probably by a quite similar mechanism), their in vivo actions may be quite different. At least one in vivo function of *E. coli* hemolysin seems to be the induction of inflammatory substances such as leucotrienes and their release from Mast cells. The hemolysin from *L. monocytogenes* (listeriolysin) allows the survival of these intracellular bacteria within the invaded target cells, but does not seem to be involved in the penetration step.

Acknowledgments. This work was supported by grants from the Deutsche Forschungsgemeinschaft (SFB 105 A-12 and SFB 165 B-4). The authors thank T. Chakraborty for many valuable discussions and E. Appel for editorial assistance.

References

1. Cavalieri SJ, Bochach GA, Snyder IS (1984) *Escherichia coli* hemolysin: Characteristics and probable role in pathogenicity. Microbiol Rev 48:326-343
2. Hacker J, Hughes C (1985) Genetics of *Escherichia coli* hemolysin. Curr Top Microbiol Immunol 116:1-162

3. Wagner W, Vogel M, Goebel W (1983) Transport of hemolysin across the outher membrane of *Escherichia coli* requires two functions. J Bacteriol 154:200-210
4. Femlee T, Pellett S, Lee EY, Welch RA (1985) *Escherichia coli* hemolysin is released extracellularly without cleavage of a signal peptid. J Bacteriol 163:88-93
5. Knapp S, Then I, Wels W, Michel G, Tschaepe H, Hacker J, Goebel W (1985) Analysis of the flanking regions from different haemolysin determinants of *Escherichia coli*. Mol Gen Genet 200:385-392
6. Hess J, Vogel M, Wels W, Goebel W (1986) Nucleotides sequences of a plasmid-encoded hemolysin determinant and its comparison with a corresponding chromosomal hemolysin sequence. FEMS Lett 34:1-11
7. Knapp S, Hacker J, Jarchau T, Goebel W (1987) Large unstable inserts in the chromosome affect the virulence properties of the uropathogenic *Escherichia coli* 06 strain 536. J Bacteriol 168:22-30
8. Hacker J, Schmidt G, Hughes C, Knapp S, Marget M, Goebel W (1985) Cloning and characterization of genes involved in production of mannose-resistant, neuraminidase-susceptible (X) fimbriae from a uropathogenic 06:K15:H31 *Escherichia coli* strain. Infect Immun 47:434-440
9. Hughes C, Hacker J, Knapp S, Duevel H, Goebel W (1987) Chromosomal deletions and rearrangements cause coordinate loss of haemolysis fimbriation and serum resistance in a uropathogenic strain of *Escherichia coli*. Microb Pathogenesis 2:227-230
10. Hacker J, Hof H, Emödy L, Goebel W (1986) Influence of cloned *Escherichia coli* hemolysin genes, S-fimbriae and serum resistance on pathogenicity in different animal models. Microb Pathogenesis 1:533-547
11. Scheffer K, Koenig W, Hacker J, Goebel W (1985) Bacterial adherence and hemolysin production from *E. coli* induces histamine and leucotriene release from various cells. Infect Immun 50:271-278
12. Koenig B, Koenig W, Scheffer J, Hacker J, Goebel W (1986) Role of *Escherichia coli* alpha-hemolysin and bacterial adherence in infection: requirement for release of inflammatory mediators from granulocytes and mast cells. Infect Immun 54:886-892
13. Felmlee T, Pellet S, Welch RA (1985) Nucleotide sequence of an *Escherichia coli* chromosomal hemolysin. J Bacteriol 163:94-105
14. Heijne G (1986) Mitochondrial targeting sequences may form amphiphilic helices. EMBO J 5:1335-1342
15. Manoil C, Beckwith J (1985) TnphoA: A transposon probe for protein export signals. Proc Natl Acad Sci USA 83:8129-8133
16. Haertlein M, Schiessl S, Wagner W, Rdest U, Kreft J, Goebel W (1983) Transport of hemolysin by *Escherichia coli*. J Cell Biochem 22:87-97
17. Smyth CJ, Duncan JL (1978) Thiol-activated (oxygen-labile) cytolysins. In: Jelaszewicz J, Wadström T (eds) Bacterial toxins and cell membranes. Academic Press, London New York, pp 129-183
18. Parrisius J, Bhakdi S, Roth M, Tranum-Jensen J, Goebel W, Seeliger HR (1986) Production of listeriolysin by beta-hemolytic strains of *Listeria monocytogenes*. Infect Immun 51:314-319
19. Kathariou S, Metz P, Hof H, Goebel W (1986) Tn916-induced mutations in the hemolysin determinant affecting virulence of *Listeria monocytogenes*. J Bacteriol 169:1291-1297
20. Kathariou S, Goebel W (1986) Introduction of the transposon Tn916 into *Listeria* ssp. Proc 9th Int Congr Listeriosis, in press
21. Mackaness GB (1962) Cellular resistance to infection. J Exp Med 116:381-406
22. Isberg RR, Falkow S (1985) A single genetic locus encoded by Yersinia pseudotuberculosis permits invasion of cultured cells by *Escherichia coli* K-12. Nature (London) 317:262-264

Iron as a Signal in Bacterial Infections

V. Braun[1], E. Fischer[1], K. Hantke[1], and H. Rotering[2]

Introduction

Insight as to how iron contributes to the pathogenicity of bacteria involves understanding the iron metabolism of the host and of the bacteria. The human host overcomes the extreme insolubility of Fe^{3+} at pH 7 by binding iron to carrier proteins, in the serum to transferrin, in secretory fluids to lactoferrin, and within cells to ferritin and hemoglobin. The serum concentration of free iron which is not bound to carrier proteins is only 10^{-18} M, which stands in sharp contrast to the iron concentration of 10^{-6} M required for bacterial growth.

The lack of readily available iron limits growth and survival of bacteria. Only those bacteria can grow which contain elaborate iron transport systems. Most, but not all, bacteria that can live on humans secrete compounds of low molecular weight (designated siderophores) with extreme affinities for Fe^{3+}. The stability constants of siderophores range from 10^{23} to 10^{52}. From these constants, biologically more meaningful pM values were calculated. pM is defined as $-\log [Fe(H_2O)_6{}^{3+}]$ when the Fe^{3+} concentration is 1 µM, and the siderophore concentration is 10 µM, at pH 7.4. For the siderophores aerobactin, coprogen, enterochelin (enterobactin), and ferrichrome, discussed in this chapter, the pM values are 23.3, 27.5, 35.5, and 25.2, respectively (1). Despite the secretion of siderophores and uptake of the ferric-siderophores via highly efficient transport systems, iron remains a scarce nutrient.

Iron is an essential constituent of the membrane-bound electron transport chains and of cytoplasmic redox enzymes. One bacterial cell contains between 10^5 to 10^6 iron ions (2). *Escherichia coli*, grown in serum, can reach densities in the order of 10^8 cells per ml. At such cell densities, 10^{14} Fe^{3+} are taken up per generation but only 10^3 free iron ions are available. Therefore, siderophores have to mobilize the transferrin-bound Fe^{3+}. On the other hand, further reduction of the iron concentration by the infected host creates bacteriostatic conditions, and gives time for the defense system to cope with an infection. The available iron is lowered by a decreased intestinal absorption of exogenous iron, deposition of iron in intracellular storage compartments, stationing of iron-binding proteins at sites of infection, and increased synthesis of host iron-binding proteins (3).

Iron deprivation of invading bacteria is convincingly demonstrated by the synthesis of certain toxins which are only formed at low iron concentration. Such proteins are the diphtheria, tetanus and Shiga toxins, toxin A, elastase, alkaline protease and hemagglutinin of *Pseudomonas aeruginosa*, cytolysins from *Clostridium perfringens*, *Listeria monocytogenes*, *Serratia*, some strains of *E. coli*, and a toxin of *Legionella* (4). Most of

[1]Mikrobiologie II, Universität D-7400 Tübingen, FRG

[2]Immuno AG, Vienna, Austria

38. Colloquium Mosbach 1987
Molecular Basis of Viral and Microbial Pathogenesis

these toxins damage cells and provide the bacteria access to intracellular iron pools. However, this assumption has not yet been proven. Iron could also serve as an environmental signal which induces not only derepression of iron transport systems, but a more general adaptation of bacterial metabolisms and of cell surface components to the site of infection in the human host. Derepressed iron transport systems have been demonstrated in *E. coli* isolated from human urine without further subculturing, and in a number of experimental animal systems. A convenient way to determine derepression of iron transport systems is the elctrophoretic separation of outer membrane proteins on polyacrylamide gels. *E. coli* expresses up to seven outer membrane proteins under iron-limited growth conditions. Five of these proteins were related to five different iron transport systems (5,6). The results obtained for *E. coli* also apply to the related Enterobacteria. It is assumed that outer membrane proteins which are snythesized in low-iron media by *Vibrio*, *Yersinia*, *Neisseria* and *Pseudomonas* strains are also constituents of iron transport systems.

The Iron Transport Systems of *Escherichia coli*

The advancement in the knowledge of transport systems relied heavily on the application of modern genetics. For this reason *E. coli* was the organism of choice. The iron (III) transport systems are composed of several distinct proteins which interact in the transport process (Table 1). Isolation and characterization of mutants provided the number of genes involved in one particular transport system. Assignment of polypeptides to the genes required cloning of the DNA fragments. This was done using multicopy plasmid vectors since the normal concentration of the transport proteins in the membrane is usually too low to be differentiated from other proteins by staining or radiolabeling on polyacrylamide gels after electrophoretic separation. The experimental conditions have to be chosen such that the transport genes are preferentially transcribed and translated in order to identify the proteins. The low amounts, their interaction, and their vectorial function prevents the isolation of single proteins and the study of their function in aqueous solution. In addition, many transport proteins are integral membrane proteins and notoriously difficult to solubilize in an active state. Their over-expression by genetic means frequently leads to cell death. Apparently, the membranes tolerate only limited amounts of a protein. Furthermore, the export machinery has a limited capacity and some over-expressed membrane proteins precipitate in the cytoplasm (16). Therefore, the expression vectors employed contained controllable promoters.

The known iron transport systems of *E. coli* (see Table 1) employ siderophores of the hydroxamate and the phenolate type. From the hydroxamates (aerobactin, ferrichrome, and coprogen) only aerobactin is synthesized by *E. coli*. The incidence of occurrence of the aerobactin transport system in natural isolates from human and animal sources is over 40% (16). Most frequently the four *aer* genes, encoding four enzymes of aerobactin synthesis and the *iut*A gene which determines the outer membrane receptor protein, are located on large self-transmissable plasmids (pCoIV), but they are also found on the chromosome (17,18). The genes are arranged and transcribed in the order *aer*D *aer*B *aer*C *aer*A *iut*A (7-9). This operon is complemented by a second operon consisting of the genes *fhu*A *fhu*C *fhu*D *fhu*B (13,19,20,21). The latter three genes are part of the aerobactin, the ferrichrome, and the coprogen transport system (Fig. 1). Although ferrichrome and coprogen are not synthesized by *E. coli*, or any other bacteria, but rather by fungi, *E. coli* transports the iron complexes via transport systems which are specified by the

Table 1. Iron (III) transport systems of *Escherichia coli*

Siderophore	Source	Proteins in Outer membrane	Periplasm	Cytoplasmic membrane
Aerobactin (617)	*E. coli* *aer*A (*iuc*D) *aer*B (*iuc*B) *aer*C (*iuc*C) *aer*D (*iuc*A)	IutA (77,345)	---	FhuB (70,329) FhuC (29,255) FhuD (28,000) TonB (26,000) ExbB (26,000)
Ferrichrome (741)	Fungi, e.g., *Ustilago*	FhuA (78,995)	---	FhuB FhuC FhuD TonB ExbB
Coprogen (821)	Fungi, e.g., *Neurospora*	FhuE (76,000)	---	FhuB FhuC FhuD TonB ExbB
Enterochelin (719)	*E. coli* *ent*A *ent*B *ent*C *ent*D *ent*E *ent*F *ent*G	FepA (79,908)	FepB (34,000)	FepC (30,500) TonB ExbB
Dicitrate (434)	Environment	FecA (80,000)	FecB (27,000)	FecC (26,000) TonB ExbB

The molecular weights given in brackets below the siderophores are those of the iron compounds.
The genes presented under "Source" designate four aerobactin biosynthesis genes termed *aer* (7) or *iuc* (8,9), and 7 enterochelin biosynthesis genes (10). The proteins bear the names of the structural genes. The exact molecular weights were calculated from the DNA sequences, the approximate molecular weights were deduced from the electrophoretic mobility on polyacrylamide gels relative to standard proteins. The *fhu*D and *ton*B genes are sequenced (13,14) but the translation initiation sites are unclear. The exact number of genes involved in the Fe^{3+}-enterochelin (11) and Fe^{3+}-dicitrate transport systems ((12), U. Preßler, H. Staudenmaier, L. Zimmermann) is not known. The transport and biosynthesis genes are regulated by a protein (16,795) encoded by the *fur* gene (15) which acts as a repressor when loaded with Fe^{2+}.

outer membrane receptor proteins (Fig. 1). The unusual participation of outer membrane proteins in single transport processes probably comes from the size and the low concentration of iron(III)-siderophores. Normally, substrates flow though water-filled channels formed by proteins, called porins, in the outer membrane. These very abundant channels, in the order of 50,000 per cell, render the outer membrane permeable for hydrophilic compounds up to a molecular weight of about 600. The iron(III)-siderophores are somewhat larger (see Table 1) so that a sufficient diffusion rate through the porin channels would require rather high concentrations in the growth medium. The outer membrane barrier and the low amount of the iron compounds is overcome by binding to outer membrane receptor proteins with apparent K_m values of 0.1 µM (22). Then the problem arises how the iron compounds are transported across the outer membrane in a vectorial manner. Dissociation from the

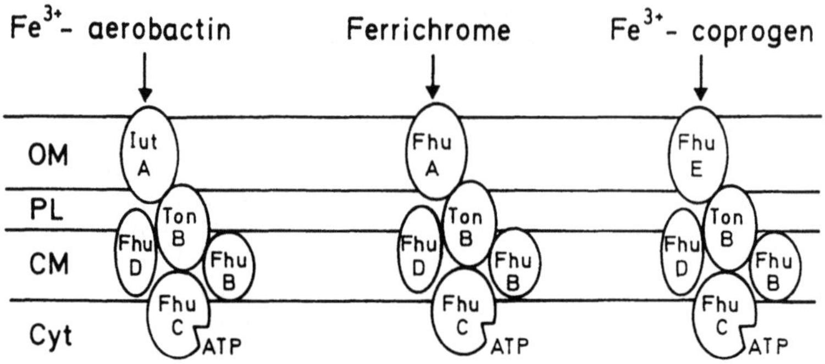

Fig. 1. Model of the Fe^{3+}-hydroxymate transport systems expressed in *E. coli*. Each transport system is composed of a receptor protein in the outer membrane *(OM)* which is specific for an iron complex of the siderophores aerobactin (*Iut*A), ferrichrome (*Fhu*A) and coprogen (*Fhu*E). The subsequent transport steps comprise four proteins common to all three transport systems. These proteins were found in the membrane fraction and were dissolved under conditions which solubilize the cytoplasmic but not the outer membrane. The proteins were arranged according to the presumptive ATP binding site of *Fhu*C (see Fig. 4), the extremely hydrophobic nature of *Fhu*B, and the proposed intermembrane bridge formed by *Ton*B. Further data on the proteins are listed in Table 1

receptor protein should involve a conformational change of the receptor proteins. Indeed, different conformational (functional) states of the *Fhu*A receptor protein were found. After binding of ferrichrome to the *Fhu*A receptor, cleavage of the protein by proteinase K proceeds at a much lower rate (E. Fischer), supporting the less pronounced lower hydrolysis rate by trypsin noticed previously (16). Ferrichrome remains at the receptor unless cells metabolize and express the *Ton*B protein. It inhibits adsorption of phage T5 to the same receptor only as long as the *Fhu*A protein is not coupled via the *Ton*B protein to cellular energy, for example in energy-starved cells, in *Ton*B mutants, in isolated outer membranes, or when isolated in pure form (23). Ferrichrome does not inhibit T5 adsorption in energized *ton*B$^+$ cells even when applied in a large surplus (23). The presence of ferrichrome at the *Fhu*A receptor under these conditions can be inferred from inhibition of phage T1 which exhibits adsorption properties different to phage T5. T1 only binds to an "energized" *Fhu*A protein in *Ton*B$^+$ cells. A simplified model suggesting different conformational states of the *Fhu*A protein is presented in Fig. 2. This model proposes a functional coupling of the *Fhu*A protein in the outer membrane via the *Ton*B protein to the cytoplasmic membrane. In response to the energized state of the cytoplasmic membrane, the *Ton*B protein assumes different conformations. The *Ton*B protein allosterically regulates the activity of the *Fhu*A and the outer membrane receptor proteins by inducing different conformations depending on its own conformation. In this view, the *Ton*B protein serves as an intermembrane protein that couples certain inner and outer membrane functions. It transmits the metabolic state of the cell to the outer membrane which otherwise is disconnected from the metabolism. In support of this hypothesis is the finding of a consensus pentapeptide in all outer membrane receptor proteins whose function depend on the *Ton*B protein (Fig. 3). The sequence of these proteins is otherwise very different. The same consensus pentapeptide has been found in colicins (bacterial toxins which kill closely related bacteria), but only in those colicins which are taken up in a *Ton*B-dependent process (27,28). A receptor mutant, *btu*B 451, with a *Ton*B$^-$ phenotype, contains a leucine to proline exchange in the consensus

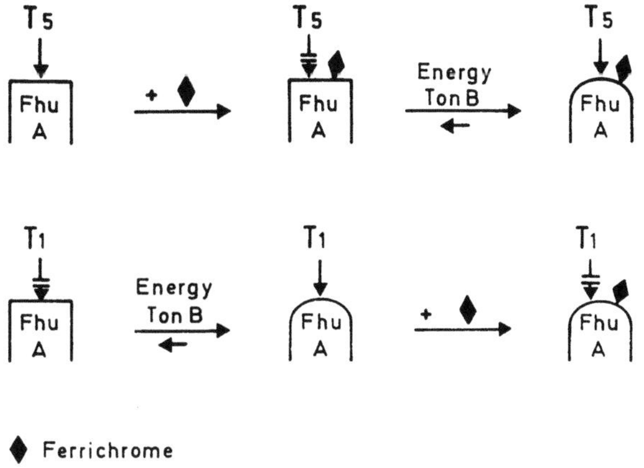

Fig. 2. Binding of phages T5 and T1 to the FhuA outer membrane receptor protein of *E. coli*. The binding of T5 is independent of energy and the TonB protein. Inhibition of T5 binding of ferrichrome occurs only in unenergized $TonB^+$ cells or in $TonB^-$ mutants *(interrupted arrow)* (23). In energized $TonB^+$ cells high concentrations of ferrichrome are required to partially inhibit T5 binding. It is proposed that FuhA assumes at least two different conformations symbolized by the rectangular, respectively arched shape. In contrast, T1 only binds to energized $TonB^+$ cells which can be inhibited by ferrichrome. The model simplifies the real situation since the contribution of the ExbB protein has been neglected (24,25). Furthermore, mutants in the *fhu*B, *fhu*C, or *fhu*D genes cannot be protected from T5 infection by ferrichrome (E. Fischer, unpublished), suggesting the involvement of the FhuB,C,D proteins in the equilibration between the two FhuA conformations

FhuE	(6)	Glu	Thr	Val	Ile	Val
FhuA	(7)	Asp	Thr	Ile	Thr	Val
BtuB	(6)	Asp	Thr	Leu	Val	Val
Iut	(6)	Glu	Thr	Phe	Val	Val
FepA	(12)	Asp	Thr	Ile	Val	Val
Colicin B	(16)	Asp	Thr	Met	Val	Val
Colicin M	(2)	Glu	Thr	Leu	Thr	Val
Colicin I	(23)	Glu	Ile	Met	Ala	Val

Fig. 3. Pentapeptide consensus sequence in the mature outer membrane receptor proteins FhuE, FhuA, Iut(A), and FepA of iron transport systems, BtuB of the mechanistically similar vitamin B_{12} transport system (26), and in colicins which are taken up into cells by a receptor and TonB-dependent process (27,28). The figures in brackets indicate the position of the first amino acid from the amino terminus

pentapeptide (26). A phenotypic revertant is not mutated back to leucine nor mutated elsewhere in the receptor protein but bears a mutation in the *ton*B gene (K. Heller, unpublished). The suppressor mutation leads to an altered TonB protein which is able to interact again with the

PROTEIN		RESIDUE	CONSERVED SEQUENCE
FhuC		30	L S L T F P A G K V T G L I G H N G S G K S T L - L K M L G R H
BtuD		19	L S G E V R A G E I L H L V G P N G A G K S T L - L A R M A G M
HisP		25	V S L Q A R A G D V I S I I G S S G S G K S T F - L R C I N F L
MalK		22	I N L D I H E G E F V V F V G P S G C G K S T L - L R M I A G L
PstB		29	I N L D I A K N Q V T A F I G P S G C G K S T L - L R T F N K M
OppD1	(Ec)	43	V T L R L Y E G E T L G V V G E S G C G K S T F - A R A I I G L
OppD2	(Ec)	42	L N F S L R A G E T L G I V G E S G S G K S Q T - A F A L M G L
OppD	(St)	40	L N F T L R A G E T L G I V G E S G S G K S Q S R L R - L M G L
RbsA	(N)	23	A A L N V Y P G R V M A L V G E N G A G K S T M - M K V L T G I
FhuC		141	L S G G E R Q R A W I A M - L V A Q D S R - - - - - - C L L L D E P T S A L D I A H Q V D V L S L V H R
BtuD		127	L S G G E W Q R V R L A A V V L Q I T P Q A N P A G Q L L L L D E P M N S L D V A Q Q S A L D K I L S A
HisP		154	L S G G Q Q Q R V S I A R A L - A M E P D - - - - - - V L L F D E P T S A L D P E L V G E V L R I M Q Q
MalK		134	L S G G Q R Q R V A I G R T L V A - E P S - - - - - - V F L L D E P L S N L D A A L R V Q M R I E I S R
PstB		152	L S G G Q Q Q R L C I A R G I - A I R P E - - - - - - V L L L D E P C S A L D P I S T G R I E E L I T E
OppD1	(Ec)	165	F S G G Q C Q R I G I A R A L - I L E P K - - - - - - L I I C D E P V S A L D V S I Q A Q V V N L L Q Q
OppD2	(Ec)	169	F S G G M R Q R V M I A M A L L - C R P K - - - - - - L L I A D E P T T A L D V T V Q A Q I M T L L N E
OppD	(St)	167	F S G G M R Q R V M I A M A L L - C R P K - - - - - - L L I A D E P T T A L D V T V Q A Q I M T L L N E
RbsA	(N)	144	L S T G D Q Q M V E I A K V L S F - E S K - - - - - - V I I M D E P T D A L T D T E T E S L F R V I R E

Fig. 4. Sequences of the FhuC protein which are homologous to proteins of binding protein-dependent transport systems. ATP binding has been shown to the MalK, HisP and OppD proteins (cited in (30)). The alignment follows the proposal of Ames (29). Sites A and B are homologous to the β-subunit of F_1, F_o-ATPases (31)

mutated receptor and suppresses the functional defect. These interpretations are consistent with the experimental data, but they are not proved. However, the results convincingly demonstrate a system in which the functional and physical interaction of two adjacent membranes can be studied.

Another interesting result of the DNA sequence determination of the entire *fhu* operon (13,20,21) is the detection of amino acid sequences in the FhuC protein (13) which are also contained in the β-subunits of F_1, F_O-ATPases of pro- and eukaryotes (31). The same homologous sequences were recently found in other transport systems through which phosphate, certain sugars, and amino acids enter *E. coli* cells. They are distinct from other transport mechanisms in that so-called binding proteins located in the perimplasm are essential constituents of these transport systems (29). However, there is no evidence that any of the proteins which catalyze iron (III)-hydroxamate transport (Fig. 1) occur in the periplasm. A possible candidate fulfilling the function of a binding protein would be FhuD, which, however, was localized in the membrane fraction along with the FhuB and FhuC proteins. The FhuB protein is extremely hydrophobic along its entire length (21), and was therefore embedded into the cytoplasmic membrane (Fig. 1). If the FuhC protein binds ATP, it should be exposed to the cytoplasm (Fig. 1). The FhuC and FhuD proteins contain only small hydrophobic segments and are otherwise rather hydrophilic (13). The FhuD protein could be exposed to the periplasm, but fixed in the cytoplasmic membrane and serve as a binding protein. If this model turns out to be true, then the iron (III)-hydroxamate transport systems would follow the mechanism of the binding protein-dependent transport. Then transport across the outer membrane would be the only peculiarity of iron uptake, while transport across the cytoplasmic membrane would follow a known but ununderstood mechanism. Indeed, proteins released by osmotic shock treatment and therefore assigned to the periplasm have been determined as constituents of the Fe^{3+}-dicitrate (L. Zimmermann, unpublished) and Fe^{3+}-enterochelin (11) transport system (cf. Table 1).

Contribution of the Aerobactin and Enterochelin Iron Transport System to the Growth of *E. coli* in Serum

Enterochelin seems to be the ideal siderophore. It is snythesized and secreted at low iron concentration, the iron (III) complex has an extremely high formation constant, and it is taken up by an efficient transport system. However, in human and animal serum enterochelin does not promote growth of *E. coli* (32). Instead, aerobactin, with a much lower iron formation constant, supports growth perfectly well (32). Several factors contribute to the superiority of aerobactin over enterochelin in serum. Aerobactin delivers iron repeatedly (32), whereas an enterochelin molecule is hydrolyzed (33) after having transported a single iron (III) ion. Enterochelin has a low solubility and stability. Due to its aromatic structure it binds to serum proteins (34). Enterochelin-specific antibodies (35) and antibodies to the Fe^{3+}-enterochelin receptor protein FepA (36) were found in normal human serum. Bound to proteins of the serum and possibly also at the bacterial surface, enterochelin apparently serves as a haptene, giving rise to the formation of antibodies. Recently, we have found that strains of *E. coli* isolated from humans are more sensitive to human serum than *aro*B mutants of these strains which are unable to synthesize enterochelin. Figure 5 shows an experiment in which the percentage of killed cells during one hour depended on the serum concentration. The K311 wild-type was rapidly killed at low serum concentrations. A much higher serum concentration was required to kill the same percentage of the *aro*B mutant which, however, became sensitive again when supplied with the missing precursor shikimate.

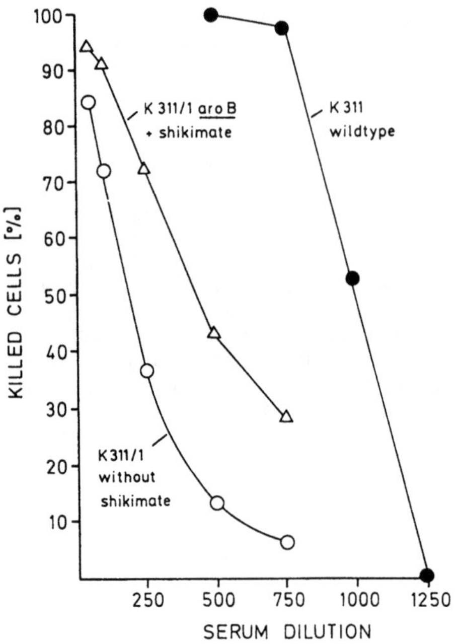

Fig. 5. Serum sensitivity of *E. coli* K311, and of the *aro*B derivative K311/1 in human serum diluted with Hanks balanced salt solution. Shikimate was added where indicated at a concentration of 0.1 mg per ml. Cells were pregrown overnight in nutrient broth containing 0.05% human transferrin which was added in iron-free form

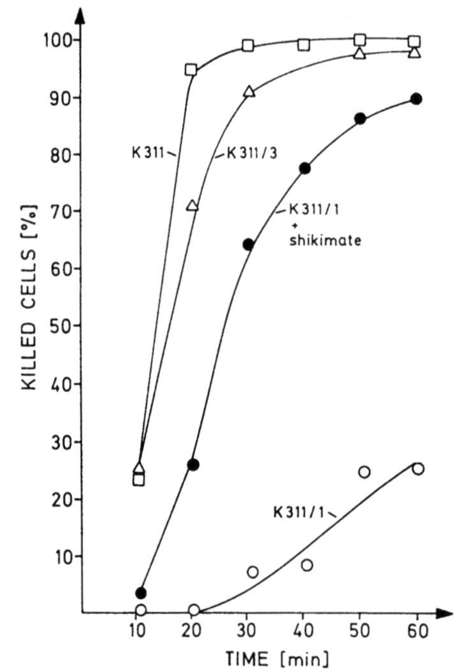

Fig. 6. Time dependence of serum sensitivity of *E. coli* K311 wild type, K311/1 *aro*B, K311/1 after addition of shikimate, the missing precursor of enterochelin synthesis, and of K311/3, an $aroB^+$ revertant of K311/1. Cells were incubated in human serum diluted 1:250 with Hanks balanced salt solution

The time dependence of the killing at a given serum concentration is shown in Fig. 6. The *aro*B mutant was virtually serum-resistant, whereas an $aroB^+$ revertant was killed nearly as rapidly as the wild-type strain. Addition of shikimate to the *aro*B mutant also resulted in a fast killing rate.

The above examples demonstrated that only a detailed knowledge of the siderophore properties and of the transport mechanisms provides an understanding of how bacteria cope with the iron limitation imposed by the host. Despite its superior iron binding, enterochelin seems to be a poor siderophore in serum. However, most enterobacteria studied to produce enterochelin so that in other environments it must be of great importance. In contrast, less than half of the *E. coli* strains from natural sources, including clinical isolates, express the aerobactin system. Moreover, the ambient in which the citrate, ferrichrome and coprogen iron transport contribute significantly to growth and survival of *E. coli* still remains to be determined.

Acknowledgments. We thank Robin Harkness for comments on the manuscript and acknowledge financial support by the Deutsche Forschungsgemeinschaft (SFB 76, SFB 323) and the Fonds der Chemischen Industrie.

References

1. Raymond KN, Müller G, Matzanke BF (1984) Top Curr Chem 123:491-102
2. Archibald F (1983) FEMS Microbiol Lett 19:29-32
3. Weinberg ED (1978) Microbiol Rev 42:45-66
4. Braun V (1985) In: Jackson GG, Thomas H (eds) Pathogenesis of bacterial infections. Springer, Berlin Heidelberg New York Tokyo, pp 168-176
5. Braun V (1985) In: Martonosi AN (ed) The enzymes of biological membranes, vol 3. Plenum, New York, pp 617-652
6. Neilands JB (1982) Annu Rev Microbiol 36:285-309
7. Gross R, Engelbrecht F, Braun V (1985) Mol Gen Genet 201:204-212
8. de Lorenzo V, Neilands JB (1986) J Bacteriol 167:350-355
9. Carbonetti NH, Williams PH (1984) Infect Immun 46:7-12
10. Flehming TP, Nahlik MS, Neilands JB, McIntosh MA (1985) Gene 34:47-54
11. Pierce JR, Earhart CF (1986) J Bacteriol 166:930-936
12. Zimmermann L, Hantke K, Braun V (1984) J Bacteriol 159:271-277
13. Burkhardt R, Braun V (1987) Mol Gen Genet 209:49-55
14. Postle K, Good RE (1983) Proc Natl Acad Sci USA 80:5235-5239
15. Schäffer S, Hantke K, Braun V (1985) Mol Gen Genet 200:110-113
16. Hoffmann H, Fischer E, Schwarz H, Braun V (1986) Arch Microbiol 145:334-341
17. Valvano MA, Crosa JH (1984) Infect Immun 46:159-167
18. Bindereif A, Neilands JB (1985) J Bacteriol 161:727-735
19. Fecker L, Braun V (1983) J Bacteriol 156:1301-1314
20. Coulton JW, Mason P, Cameron DR, Carmel G, Jean R, Rode HN (1986) J Bacteriol 165:181-192
21. Köster W, Braun V (1986) Mol Gen Genet 204:435-442
22. Wookey P, Hussein S, Braun V (1981) J Bacteriol 146:1158-1161
23. Hantke K, Braun V (1978) J Bacteriol 135:190-197
24. Hantke K, Zimmermann L (1981) FEMS Microbiol Lett 12:31-35
25. Eick-Helmerich K, Hantke K, Braun V (1987) Mol Gen Genet 206:246-251
26. Heller K, Kadner RJ (1985) J Bacteriol 161:904-908
27. Schramm E, Mende J, Braun V, Kamp RM (1987) J Bacteriol 169:3350-3357
28. Dreher R, Braun V, Wittmann-Liebold B (1985) Arch Microbiol 140:343-346
29. Ames GF-L (1986) Annu Rev Biochem 55:397-425
30. Higgins CF, Hiles JD, Salmond GCP, Gill DR, Downie JA, Evans IJ, Holland JB, Gray L, Buckel SD, Bell AW, Hermodson MA (1986) Nature (London) 323:448-450
31. Ducan TM, Parsonage D, Senior AE (1984) FEBS Lett 208:1-6
32. Braun V, Brazel-Faisst C, Schneider R (1984) FEMS Microbiol Lett 21:99-103
33. Rosenberg H, Young JG (1974) In: Neilands JB (ed) Microbial iron metabolism. Academic Press, London New York, pp 67-82
34. Konopka K, Neilands JB (1984) Biochemistry 23:2122-2127
35. Moore DG, Earhart CF (1981) Infect Immun 31:631-635
36. Griffiths E, Stevenson P, Thorpe R, Char H (1985) Infect Immun 47:808-816

Clostridial Neurotoxins – The Search for a Common Mode of Action

E. Habermann[1]

Introduction

Modern pharmacology is primarily devoted to the mode of action of drugs. Conversely, a detailed knowledge of their chemistry, pharmacokinetics and pharmacodynamics will improve our understanding of many biological events. Once the latter are known, drugs and events can be held together by a central theory. With some bacterial toxins this goal has been achieved. For instance ADP-ribosylating toxins (Foster and Kinney 1985, Wreggett 1986) like diphtheria, cholera, pertussis toxin, pseudomonas exotoxin A, and also the cytolytic botulinum C2 toxin (Aktories et al. 1986) can be defined by their biologically relevant substrates. Cytolytic toxins like staphylococcal α-toxin or the thiol-activated, cholesterol-binding toxins like streptolysin-O are inserted into biomembranes, form ring-like aggregates and surround artificial pores (Bhakdi this vol.). Still others act in a detergent-like manner, like streptolysin S or staphylococcal θ-toxin (for review see Le Vine and Cuatrecasas 1986).

Clostridia, too, produce many toxins. Some enzymes, like hyaluronidases, neuraminidases, collagenases or other proteases, serve as adjuvants for bacterial growth or spreading of the actual toxins (Fig. 1). Others, like phospholipases C or members of the large group of oxygen-labile, cholesterol binding hemolysins, damage neighboring cells.

Clostridial neurotoxins are distinct from the agents mentioned. In animals their LD_{50} is in the range of a few ng/kg. Their symptoms in man are well known, and much work has been devoted to the pathogenesis of tetanus and botulism during the past 100 years. Despite many incentives and attempts (for recent review see Habermann and Dreyer 1986), we are not yet able to deduce tetanus and botulism from molecular events. Therefore this chapter presents pieces of evidence for a common structure and mode of action of the toxins, hoping that they will yield a unifying picture.

Common Biochemical Background

Neurotoxins must have been a great innovation in the chemical warfare of anaerobic bacteria. Whereas the previously named toxins damage by local attack, neurotoxins kill the host by a systemic approach, thus converting him into an anaerobic fermenter which feeds the neurotoxic bacteria for a long time. From these evolutionary aspects, it appears improbable that Nature had made the same invention twice. It may be predicted that tetanus and botulinum neurotoxins stem from one common ancestor protein.

---

[1] Buchheim-Institut für Pharmakologie, Justus-Liebig-Universität, Frankfurter Strasse 107, D-6300 Giesen, FRG

38. Colloquium Mosbach 1987
Molecular Basis of Viral and Microbial Pathogenesis

CLASS	REPRESENTATIVE	MAIN PRODUCER
CYTOTOXIC ENZYMES	PHOSPHOLIPASE C	CL. PERFRINGENS
" HAEMOLYSINS	θ-TOXIN	CL. PERFRINGENS
	TETANOLYSIN	CL. TETANI
ADP-RIBOSYLATING ENZYMES	C_2-TOXIN	CL. BOTULINUM C_2
	C_3-TOXIN	CL. BOTULINUM C_1, C_2
		CL. DIFFICILE
	J-TOXIN	CL. PERFRINGENS
ENZYMES WITH EXTRACELLULAR SUBSTRATES	COLLAGENASE	CL. PERFRINGENS
		CL. NOVYI
		CL. HISTOLYTICUM
	HYALURONIDASE	CL. PERFRINGENS
	NEURAMINIDASE	CL. PERFRINGENS
NEUROTOXINS	BOTULINUM TOXIN A - F	CL. BOTULINUM A, B, C_1, D, E, F
	TETANUS TOXIN	CL. TETANI

Fig. 1. Clostridia as sources for toxins

At first sight the genetic representation of the neurotoxins is very variable. DNA for tetanus toxin is located in a large (75 kb) plasmid. *Cl. botulinum* C_1 and D toxins are encoded in bacteriophages, and the same appears to be true for *Cl. novyi* toxin. The bacteria can be "cured" of their genuine phages, and the C_1 phage can be replaced by the D phage, and vice versa. The phage from *Cl. novyi* can be brought into *Cl. botulinum* C_1, but not vice versa (Fig. 2). The other neurotoxins are apparently encoded in the bacterial chromosome.

PLASMID	TETANUS TOXIN
BACTERIOPHAGES	CL. BOTULINUM C_1, D
	CL. NOVYI
CHROMOSOME	CL. BOTULINUM A, B, E

Fig. 2. Gene carriers for clostridial toxins (Habermann and Dreyer 1986)

However, clostridial neurotoxins have many biochemical properties in common (Fig. 3). They are synthesized as single chains of about 150,000 Da, which contain at least one internal disulfide bridge. Upon release into the extracellular medium, they are nicked by accompanying proteases to yield a N-terminally located light chain, followed by a heavy chain. All clostridial neurotoxins are weakly acidic and contain some free SH-groups. They differ immunologically, except C_1 and D toxin, which share many antigenic sites and appear to form a subgroup. So far, no antibody recognizes antigenic determinants common to tetanus and any botulinum toxin (for review see Habermann and Dreyer 1986). It should be noted that C_2-toxin makes an exception. Its C_2I and C_2II moieties correspond to light and heavy chains in size, but are not held together neither covalently nor noncovalently. Not incidentally, C_2 toxin is not a neurotoxin but a cytotoxin, and acts by ADP-ribosylation of G-actin.

The localization of the tetanus toxin gene on a plasmid facilitated the elucidation of its primary structure (Eisel et al. 1986). Shortly,

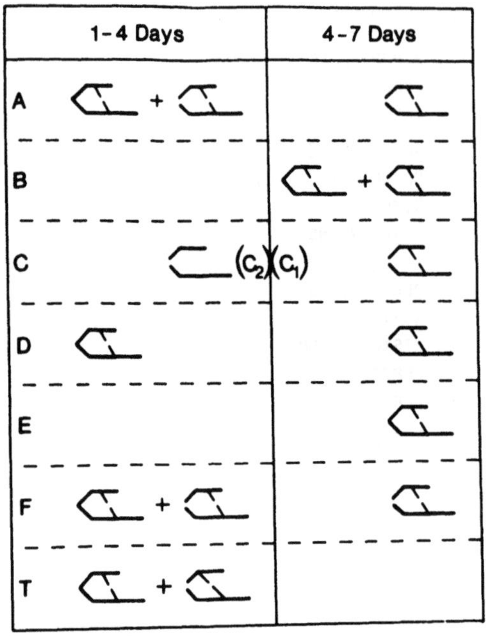

Fig. 3. Structure of individual botulinum (A-F) neurotoxins derived from bacterial cultures of varying age (DasGupta 1981). Tetanus toxin (*T*) is added for comparison

tetanus toxin was sequenced in a conventional manner at three positions: the N-terminus of the light chain which is also the start of the unnicked toxin, the N-terminus of the heavy chain (which is not strictly homogeneous), and that of the so-called fragment C which results from papain digestion (Fig. 4). Then a DNA sequence was synthesized representing some amino acids of the latter, radiolabeled, and fragments of the isolated plasmid were screened with the aid of the probe and cloned. Thus the complete structure of the toxin gene was determined. So far it does not explain the extreme toxicity of the molecule. There is no larger hydrophobic stretch which is typical of transmembranal proteins. Thus it is extremely improbable that the toxin can be converted into a transmembranal protein during its passage into the cell. Its way must be hydrophilic.

The primary structure of tetanus toxin indicates close homologies with botulinum toxins, as far as the N-terminal moieties of heavy and light chains are concerned. They have been sequenced at the protein level by some groups. The similarities are the more striking as homologies were absent in the sequences of more than 3600 other proteins. Thus the clostridial neurotoxins form a novel group of proteins (Fig. 5).

Common Pharmacological Properties of the Toxins

If we look merely at the clinical manifestations, any comparison between the toxins looks absurd. Tetanus toxin acts on the central nervous system and kills by spasticity, whereas botulinum toxins paralyze the victim and kill him by peripheral respiratory paralysis. However, evidence has accumulated that tetanus toxin, appropriately applied, can also paralyze neuromuscular junctions in situ, even in man.

Introduction of an isolated preparation, the phrenic nerve-hemidiaphragm of the mouse, into tetanus research (Habermann et al. 1980)

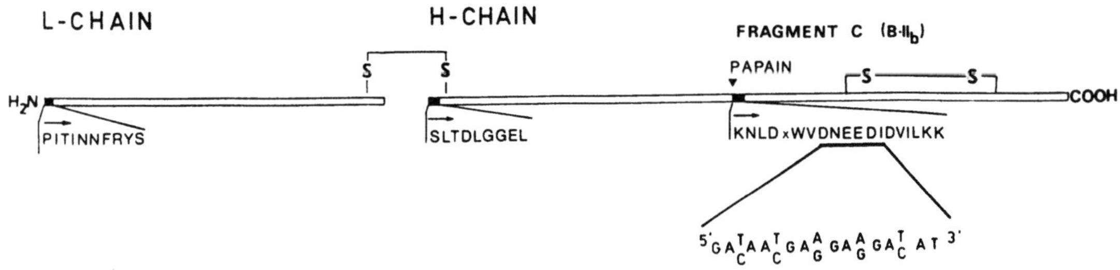

Fig. 4. Structural model of tetanus toxin and partial amino terminal sequences of individual subfragments. Partial amino acid sequences of the L- and H-chains as well as of fragments C and BII_b were determined by Edman degradation. A pool of synthetic oligonucleotides corresponding to amino acids 8-13 of fragment C was used to identify the gene encoding tetanus toxin (Eisel et al. 1986)

LIGHT CHAINS	1				5					10					15					20		
TETANUS TOXIN [a]	P	I	T	I	N	N	F	R	Y	S	D	P	V	N	N	D	T	I	I	M	M	E
BOTULINUM A [b]	P	F	V	N	K	Q	F	N	Y	K	D	P	V	N	G	V	D	.	.	.	.	.
BOTULINUM B [c]	P	V	T	I	N	N	F	N	Y	N	D	P	I	D	N	N	.	.	.	.	.	.
BOTULINUM E [c]	P	-	K	I	N	S	F	N	Y	N	D	P	V	N	D	R	T	I	L	Y	I	.

HEAVY CHAINS																										
TETANUS TOXIN [a]	S	L	T	D	L	G	G	E	L	C	I	K	I	K	N	E	D	L	T	F	I	A	E	K	N	S
BOTULINUM A [b]					A	L	N	D	L	C	I	K	V	N	N	I	D	L	K	F	.	.	.	.	.	.
BOTULINUM A [d]					A	L	N	D	L	C	I	K	V	N	N	W	D	L	F	F	S	P	S	E	D	N
BOTULINUM B [c]					A	-	P	G	I	C	I	D	V	D	N	E	D	L	F	F	I	A	D	.	.	.
BOTULINUM E [c]							K	S	I	C	I	E	I	N	N	G	E	L	-	F	.	.	.	.	.	.

Fig. 5. Structural homologies between tetanus toxin and botulinum toxins A, B and E found in the N-terminal amino acid sequences of the respective L- and H chains. Amino acids common to tetanus and botulinum neurotoxins are framed with *solid lines*. *a* Sequence from DNA analysis (Eisel et al. 1986); *b* Schmidt et al. (1984); *c* Schmidt et al. (1985); *d* Shone et al. (1985)

was helpful. At the macroscopical level, tetanus and botulinum A toxin appeared to be indistinguishable. Both toxins depressed the neuromuscular transmission by a presynaptic attack (Fig. 6). Their effect depended strongly on temperature. It was preceded by a latency period which was shortened with increasing toxin concentrations. Already after a part of the latency period had passed, progress of poisoning could no longer be stopped, and manifest paralysis was no longer resolved by washing or repeated addition of antitoxin. Tetanus toxin appeared not to differ from botulinum A toxin except by the higher (about 500-fold) dosage required for paralysis. However, some discrimination became possible by analysis of the endplate potentials which are indicative of the quantal transmitter release. As expected, botulinum A toxin diminished the number of quanta per endplate potential, but left its form and its coupling to the stimulus intact. Teta-

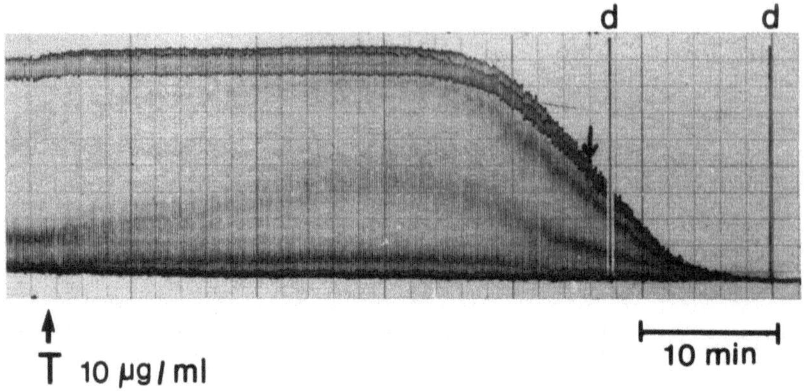

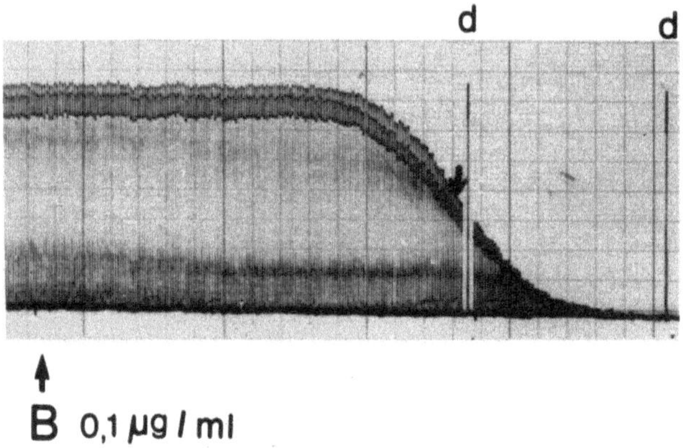

Fig. 6. Similar time course of the development of the paralysis induced by tetanus toxin and botulinum toxin. Mouse phrenic nerve-hemidiaphragms poisoned with 10 µg/ml tetanus toxin (*T*) or 0.1 µg/ml botulinum A toxin (*B*). *d* direct stimulation of the muscle. The *downward arrows* give the time $t_{0.5}$ where the contractile amplitude decreased to 50% of the initial value. The phrenic nerve was continuously stimulated at 0.1 Hz (Habermann et al. 1980)

nus toxin, however, altered the transmitter release in two aspects. First, the individual endplate potentials were dispersed on the time axis. Moreover, when a train of stimuli was applied, the toxin delayed the onset of the postsynaptic response, as if the nerve ending had to accumulate some stimulus-triggered events before it released amounts of transmitter sufficient for an endplate potential (Fig. 7). Thus, although all clostridial neurotoxins depress the neurotransmitter release, they can be subdivided into two subgroups. Botulinum A toxin is the only representative of the paralytic group, and tetanus toxin, but also botulinum B toxin, belong to the paralytic-dispersing group.

Work with the phrenic nerve-hemidiaphragm allowed to develop a three-stage concept of poisoning which applies to both botulinum (Simpson 1980) and tetanus (Schmitt et al. 1981) toxins. It is based merely on the variation of exposure times, temperature, and application of antisera, and distinguishes three steps: Binding (a) is followed by translocation (b), and steps (a + b) are the precondition of action (c). It is proposed that the hypothesis applies whenever a neurotoxin effect can be registered. A multitude of systems is necessary for a thorough analysis because the diaphragm is indispensable for electrophysiology but does not allow much biochemistry.

Binding. Studies on neural membranes have shown that both tetanus and botulinum toxins are bound to gangliosides. It is generally accepted today (see Habermann and Dreyer 1986) that binding depends largely, but not exclusively, on gangliosides. Affinity appears to be best

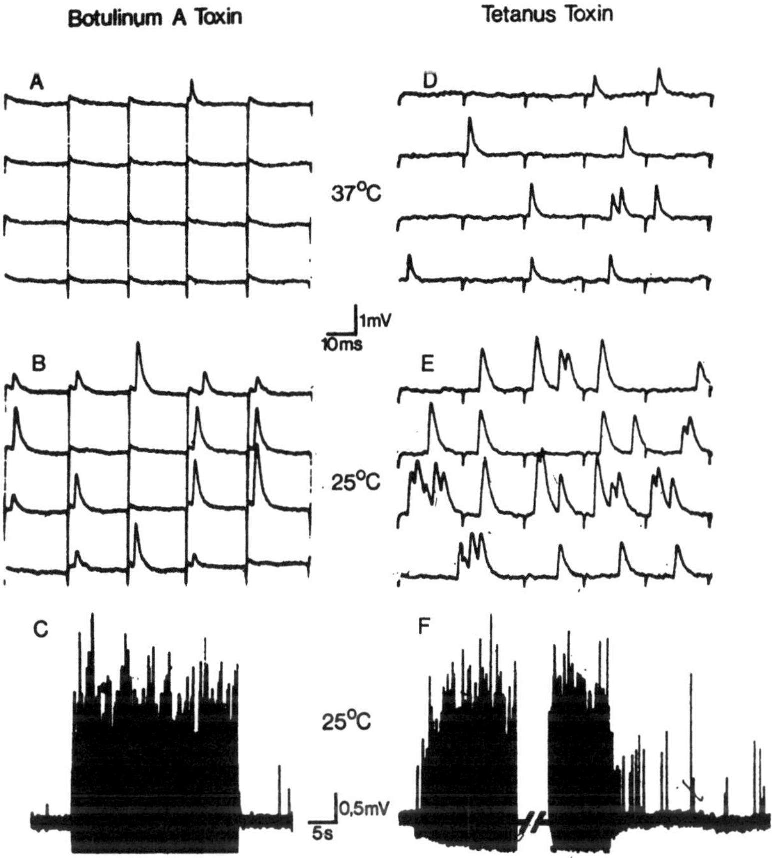

Fig. 7. The effect of temperature on transmitter release from endplates which had been treated with botulinum A (*A–C*) or tetanus (*D–F*) toxin (Dreyer and Schmitt 1983). *A*, *B*, *D* and *E* are four traces taken at random from recordings made during 50 Hz stimulation trains. The downward deflections in the voltage traces are due to the nerve stimulus artifacts. *F* shows the onset and cessation of responses to a 50 Hz train of 90 s duration. For the BoTx experiment a cut muscle preparation was used. Membrane potential was - 54 mV for botulinum A and - 62 mV for tetanus toxin

to those of the GD1b, GT1 and GQ type, however with some differences between the toxins. Binding is depressed by treatment of the membranes with *V. cholerae* neuraminidase. Some fixation of tetanus toxin persists after neuraminidase treatment. Three reasons are advocated. Either the enzymatic degradation of the gangliosides is incomplete, and some binding persists due to a large receptor excess, or the resulting GM1 binds the toxin sufficiently, or there is still another, general binding site involved. Whatever the final outcome may be, it cannot be decided whether gangliosides serve merely as pharmacokinetic receptors, i.e. initiate the toxin translocation (b) into the membrane or into the interior of the cell. Otherwise, they might function as pharmacodynamic receptors, i.e., mediate the toxin effect directly by initiating a chain of events. Without doubt, ganglioside binding is involved in poisoning by both tetanus and botulinum toxins, but its role is still unsettled.

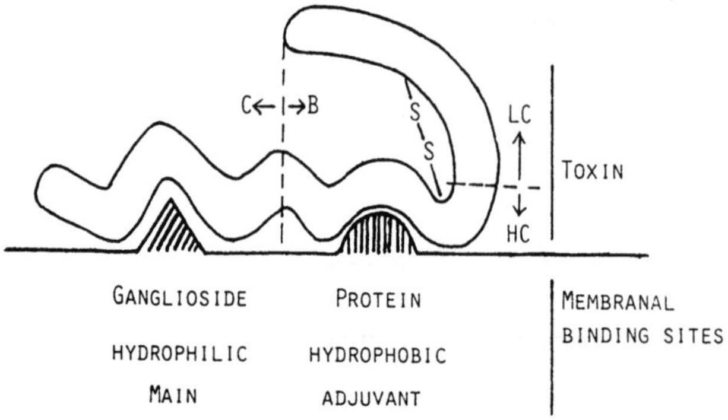

Fig. 8. Two binding sites (hypothetic) for tetanus toxin binding. *C* fragment C; *B* fragment B; *HC* heavy chain; *LC* light chain. The number of binding sites per molecule may be >2. Arguments for the model are compiled in Fig. 9

I. AT THE TOXIN LEVEL

- **FRAGMENT C BINDS TO GANGLIOSIDES SPECIFICALLY**
- **FRAGMENT B DOES NOT BIND TO MEMBRANES**
- **THE COMPLETE TOXIN IS REQUIRED FOR BINDING TO BRAIN, BUT NOT TO KIDNEY MEMBRANES IN HIGH IONIC STRENGTH BUFFER**
- **ASCENT TO SPINAL CORD IS BETTER WITH COMPLETE TOXIN AS COMPARED WITH FRAGMENT C**
- **NICKED AND UNNICKED TOXIN BIND EQUALLY.**

II. AT THE MEMBRANE LEVEL

- **ONLY THOSE MEMBRANES WHICH CONTAIN LONG-CHAIN GANGLIOSIDES BIND TETANUS TOXIN**
- **BOTH NEURAMINIDASE AND PROTEASES DEPRESS BINDING OF TOXIN**
- **BINDING IS MUCH DEPRESSED BY HEATING TO 56° C**
- **THE HEAVY CHAIN OF TETANUS TOXIN FORMS PORES IN ARTEFICIAL MEMBRANES.**

Fig. 9. Evidence for cooperativities in tetanus toxin binding (references on p. 165)

There is some evidence that, besides gangliosides, a heat- and protease-sensitive cell constituent participates in toxin binding (Figs. 8,9). The binary character of the membranal binding site(s) might be mirrored by the complex binding behavior of tetanus toxin (Yavin and Nathan 1986; Habermann and Albus 1986; Lazarovici and Yavin 1986; Pierce et al. 1986; Montecucco 1986).

As mentioned (Fig. 4), the toxins can be split enzymatically into a single-chain fragment C which represents the C-terminal (MW 50,000) half of the heavy chain, and the bichainal fragment B that consists of the N-terminal remainder of the heavy chain, together with the light chain (MW 100,000). Only the C-fragment interacts with gangliosides, and it does not change the electrical resistance of arteficial membranes. Fragment B, but not its light chain moiety, forms pores on black lipid membranes and increases the K^+ permeability of asolecitin vesicles, probably by way of hydrophobic interactions (Roa and Boqüet 1985). Thus the assumption is justified that both parts of the heavy chain are involved in binding and translocation. It is strengthened by recent data (Weller et al. 1986) on binding to brain membranes and ascent in rats (see below). Even in the binding assay, forces outside fragment C contribute to binding of the complete tetanus toxin molecule. They can be subtracted when the assay is run in high salt buffer.

The second receptor on the N-terminal moiety of the heavy chain is apparently hydrophobic, as evident from its protection against proteolysis by contact with asolectin vesicles (Roa and Boquet 1985). The hydrophobic site appears to be relevant for both binding and translocation. So far, knowledge of the primary structure of tetanus toxin has not proven any longer hydrophobic stretch in the expected position. The negative finding militates against the assumption that tetanus toxin may enter membranes as an integral membrane protein. Clustering of short hydrophobic stretches may nevertheless enable binding to hydrophobic membranal counterparts; as to botulinum toxins, there are also some evidences for nongangliosidic attachment sites (see Habermann and Dreyer 1986).

Translocation and Axonal Transport. The usual autoradiographic techniques do not allow a clear distinction between binding to membranal surfaces, translocation into the membranes, and internalization into the tiny cytoplasmatic space of nerve endings or dendrites. Tetanus toxin loses some accessibility to pronase upon prolonged contact with nerve cell cultures (Staub et al. 1986). Internalization of botulinum A toxin has been shown by electron microscope autoradiography of the neuromuscular junction (Dolly et al. 1984). However, data on the way and mode of translocation of tetanus and botulinum toxins into living systems are rare. Work on artificial membranes may serve as an alternative. When exposed to tetanus (Hoch et al. 1985) or botulinum C (Donovan and Middlebrook 1986) toxin in the acidic range, pores of about 10 pS in size are formed in black lipid membranes. Gangliosides are helpful, but not essential to mediate the toxin effects. It is hypothesized that low pH mimics the environment in intracellular compartments, and pore formation might be regarded as an equivalent of translocation. The process implies both binding to a nongangliosidic site and passive transport.

Translocation is a precondition of axonal transport. Conversely, axonal transport is the ultimate proof for the preceding translocation, which is, at least in this case, an internalization. To show the common pharmacokinetic behavior of tetanus and botulinum A toxin, they were radiolabeled and injected into the m. gastrocnemius of rats. The spinal cords were processed for autoradiography about 1 day later. Both neurotoxins were concentrated within the ventral horn, where the representative motoneurones are known to be located (Fig. 10). Ascent of botulinum toxin was less but still pronounced (Fig. 11). According to the concept of binary binding sites on both toxin and membranes, fragment C ascends to the spinal cord in rats very much less than the complete toxin.

At least in the case of tetanus toxin, axonal transport is followed by transsynaptic transport from the spinal motoneurones to the inhibitory interneurones. This sequence is crucial for the spinal effects of tetanus toxin. It is unknown why botulinum A toxin fails to act centrally, although it ascends and is active on isolated neural systems. In this context it must be considered that, due to the much higher potency of botulinum A toxin at the peripheral neuromuscular junction, the animal will not survive the time required for a sufficient ascent. Moreover, we do not know whether the spinal radioactivity indeed still represents botulinum toxin. The question has been answered, in a partially positive sense, for tetanus toxin only (Habermann et al. 1977).

Action. Analysis of steps (a) and (b) merely deals with the preconditions of action, which still occurs in a black box. A multitude of approaches has been directed towards step (c), so far with partial

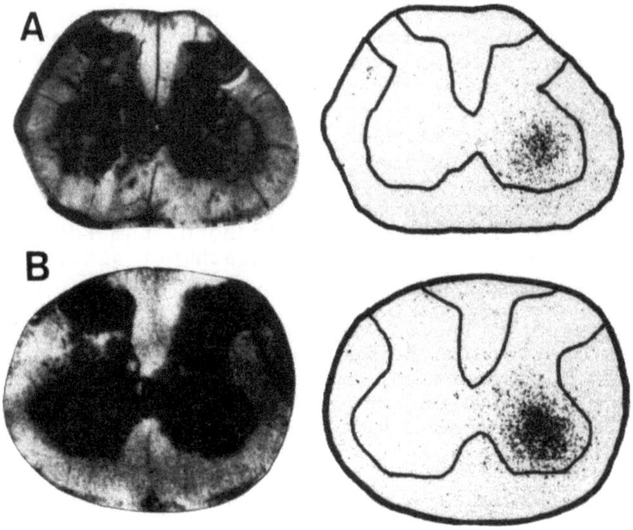

Fig. 10A,B. Distribution of radioactivity in the various areas of rat spinal cord after injection of ^{125}I-labeled tetanus (*B*) or botulinum A (*A*) toxin into the ipsilateral gastrocnemius muscle of rats. The *left part* depicts the preferential staining with methylene blue of the grey substance. The *right part* represents the adjacent autoradiography. (Weller and Habermann, to be published)

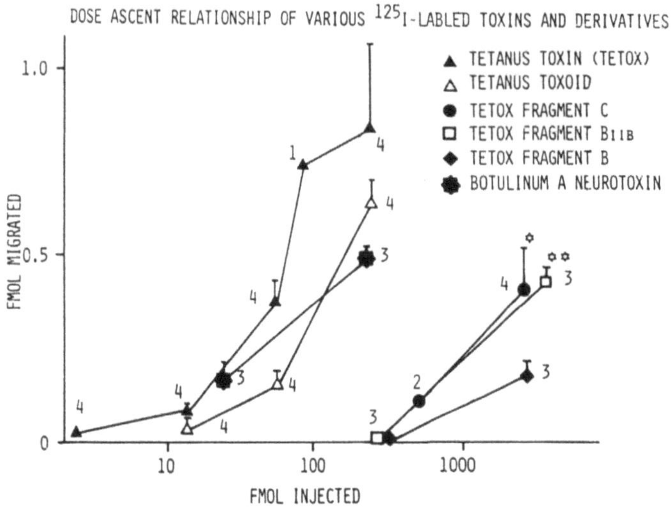

Fig. 11. Relationship between the amount of radiolabeled toxin and some derivatives injected and their accumulation in the rat spinal cord. (Weller et al. 1986). The *ordinate* shows the amount of tracer in the lumbar part of the spinal cord, calculated from the respective specific radioactivity. The abscissa gives the amount injected. 125-albumin did not measurably ascend, even when 2500 fmol were injected. The *number of animals* is given at the data points. The results are expressed as mean values ± S.D. Fragments C and BIIb differ from fragment B at the levels of $^{*}P<0.05$ and $^{**}P<0.01$ with respect to their ascent

success only. However, many functional equivalents between tetanus and botulinum toxins have become apparant. We have already mentioned the neuromuscular synapse as their common target, and some differences in their potency and the quality of their effects. Some other isolated systems turned out to be equally informative, provided they dealt with neurotransmitter release either directly or indirectly. In particular, broken or cultured nerve cells were helpful, the for-

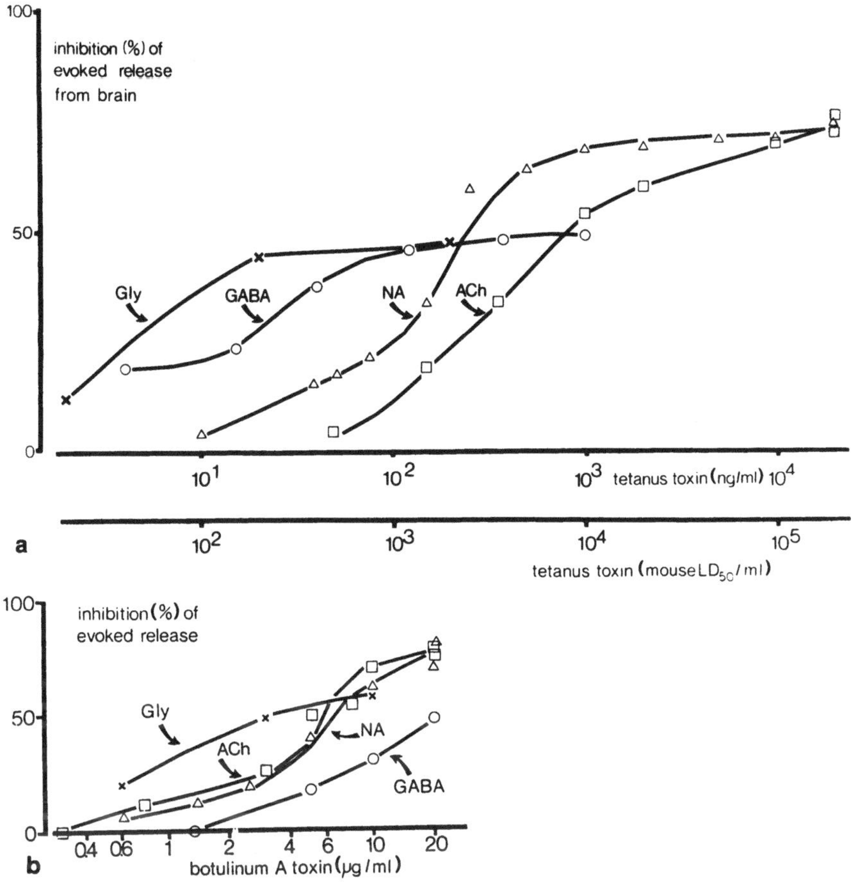

Fig. 12. *a* Inhibition by tetanus toxin of the release of various transmitters evoked by high K^+ from superfused broken cell preparations. *Ordinate*: Inhibition (%) of evoked release. *Abscissa:* Concentration of tetanus toxin in ng/ml (*upper scale*) and mouse LD_{50}/ml (*lower scale*). Inhibition was calculated using the quotient of the means of three poisoned and three unpoisoned samples. Glycine (x—x), GABA (O—O)), acetylcholine (□—□), noradrenaline (Δ—Δ). Preparations from forebran (glycine, GABA and acetylcholine) and striatum (noradrenaline) were used. *b* Inhibition by botulinum A toxin of the evoked release from broken cell preparations of the central nervous system. Preparations from forebrain were preloaded with choline or GABA, from spinal cord with glycine, from striatum with noradrenaline. The *abscissa* gives the toxin concentration (µg/ml). One µg/ml is equivalent to about 10^4 mouse LD_{50}/ml (Bigalke et al. 1981b)

mer by their convenience, the latter by their longevity. Both were susceptible to tetanus as well as botulinum A toxin.

Studies on Broken Cell Preparations. We have assessed glycine, GABA, noradrenaline, acetylcholine, met-enkephalin-like material and d-aspartate, which is equivalent to glutamate as an excitatory amino acid (Bigalke et al. 1981a,b, Albus and Habermann 1983, Habermann 1981, Janicki and Habermann 1983). Both tetanus and botulinum A toxin inhibited the release of every transmitter investigated (Fig. 12a,b). Like on the mouse diaphragm, their effect depended on time and temperature. It was apparent after a latency period. Once manifest, toxicity was no longer reversed by antitoxin or washing. Whereas on the diaphragm tetanus toxin was very much less potent than botulinum A toxin, the reverse was true with broken brain cells. Here the requirement for botulinum A toxin was high, and depended not much on the transmitter studied. Tetanus toxin, however, was most potent against the release of inhibitory transmitters, as expected from clinical tetanus.

Primary Neural Cultures. Long incubation periods are required with clostridial neurotoxins, while the viability of broken cell preparations decreases. Thus introduction of primary cell cultures to study neurotransmitter release (Bigalke et al. 1978) was advantageous. They can be kept for months, are well accessible to the high molecular weight toxins, and can be impaled with microelectrodes. They are excitable and spontaneously active. When poisoned with tetanus toxin in doses down to 10^{-11}M and less, they initially pass a latency period. Then they convert their random firing pattern into paroxysmal depolarizing events (PDE), and finally they cease to fire for weeks (Fig. 13). The cells remain viable and excitable upon direct electrical stimulation. The background of the dramatic alteration becomes evident by counting the excitatory and inhibitory p.s.p's at different times after poisoning. When the PDE's commence, the i.p.s.p.'s decrease until they disappear. Together with the PDE's, the e.p.s.p.'s are also lost. Thus we understand the different sensitivity of the excitatory and inhibitory synapses as the physiological basis for tetanus at least in cultures.

Botulinum A toxin differs in quantitative terms only. Like on broken brain cell preparations, the toxin is considerably less potent than tetanus toxin on primary nerve cell cultures, and it apparently does not discriminate between excitatory and inhibitory synapses. Only a few cells develop the PDE's so typical of tetanus toxin. Others are paralyzed without preceding excitation, some are not influenced at all.

If we correlate the dose-response curves for the different synapses mentioned, we can subdivide them into two groups. Of high sensitivity are the neuromuscular junctions for botulinum toxin, and the inhibitory neuronal synapses for tetanus toxin. Of low sensitivity are the neuromuscular junction for tetanus toxin, the excitatory neuro-neuronal synapses for tetanus toxin and all neuro-neuronal synapses for botulinum A toxin (Bigalke et al. 1985). In contrast to tetanus toxin, the effect of botulinum A toxin is neuraminidase-sensitive (Bigalke et al. 1986).

The data mentioned for cell cultures apply to electrophysiology only. Neurochemical studies are just in their beginnings. Primary cultures from embryonic rat brain were used for K^+-evoked acetylcholine release. It was inhibited by both tetanus and botulinum A toxin, the latter being about 1000 times more potent (Bigalke et al. 1978).

Due to their longevity, primary cultures can also be used to study the fate of tetanus toxin. Previously (Habermann et al. 1977), it had been shown in rats and cats that tetanus toxin persists for many days within the spinal cord, apparently with only moderate degradation. This view has been extended and confirmed on primary nerve cell cultures. The half-life time of intact toxin and its both chains were between 4-5 days. The toxin could not be removed by washing and antibodies. The persistence of the toxin may well account for the long duration of tetanus both in vivo and in vitro. Moreover it may be concluded that the toxin, unlike diphtheria or cholera toxin, is not measurably processed before its entrance into or within the cell (Habig et al. 1986). Nothing is known about the persistence of botulinum toxin.

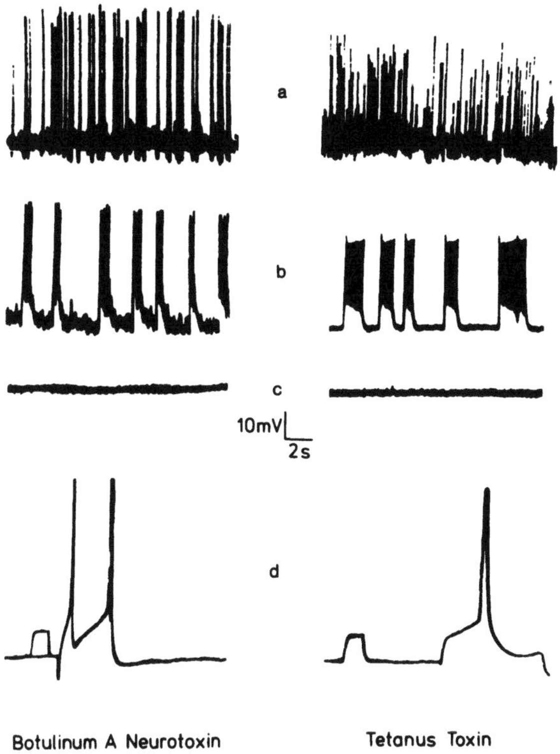

Fig. 13a-d. Paroxysmal depolarizing events followed by blockade of spontaneous synaptic activity due to botulinum A neurotoxin (7×10^{-9}M) and tetanus toxin (3×10^{-10}M). *a* Control recordings show the spontaneous synaptic activity obtained from two different nerve cells immediately after the application of the toxins at 37°C. *b* The toxins produced paroxysmal depolarizing events after a latent period of about 60 min. The membrane potential depolarized by about 20 mV for 500 ms to several seconds, eliciting a firing of action potentials. In the case of botulinum A neurotoxin they were still accompanied by continuous synaptic activity, whereas no synaptic potentials occurred during the interburst-intervals in cells treated with tetanus toxin. *c* The burst like activity was followed by a complete blockade of postsynaptic activity, for botulinum A neurotoxin after about 200 min and for tetanus toxin after about 500 min. *d* Direct stimulation by a current pulse (1 nA) elicited action potentials in cells blocked by tetanus toxin or botulinum A neurotoxin. Membrane potentials were not altered by the toxin treatment (-62mV). Calibration pulse: 10 mV, 10 ms (Bigalke et al. 1986)

The Search for a Common Primary Event

What is the primary action of clostridial neurotoxins, and which parts of the molecule are responsible? As to the first question, we can only say which processes we have found to be *not* sensitive to tetanus and botulinum A toxin: Ca^{2+} entry, production of cAMP, of cGMP, of inositol phosphates, ion-dependent GABA transport, resting potential of the pericaryon, actin polymerization, microtubule assembly. Moreover, tetanus toxin neither ADP ribosylates nor phosphorylates any known substrate. Mellanby (1984) presumes that Ca^{2+} sensitivity of a factor involved in exocytosis is depressed. However, we can only state that the action of the toxin occurs anywhere downstream Ca^{2+} entry.

We (Aktories et al. 1986) have found that C_2-toxin ADP-ribosylates actin. Aktories et al. (1987) have described another ADP-ribosylating enzyme produced by a strain of Cl. botulinum C and called it C_3. Its substrate is a 21,000 Da protein. Both enzymes are not neurotoxins, i.e., not specific for nerve cells. Ohashi et al. (1987) reported that botulinum C and the related D neurotoxins, however in high concentrations, also ADP ribosylate a 21,000 Da substrate. Aktories (to be published) has convincing evidences that his novel C_3 enzyme is responsible. Thus the molecular mode of action of the clostridial neurotoxins still waits for elucidation.

The second question has been brought closer to answer by recent work in Neher's laboratory (Penner et al. 1986). To circumvent the pharmacokinetic receptor, the agents were injected into cultured adrenal medullary cells. One hour later Ca^{2+} was injected and the evoked exocytosis measured by the patch clamp technique. Both tetanus and botulinum A toxin prevented the response, which indicates that their target is not restricted to neuronal cells provided they gain access to the cytosol (Fig. 14). With other words, they are neurotoxins under superficial aspects only. Now the way was open to test which part of tetanus toxin was responsible. Fragment C turned out to be completely inactive, whereas fragment B mimicked the effect of the complete toxin. Thus fragment B, which exerts hydrophobic binding, may be involved in translocation, and is pharmacologically active, once it has been brought on the right place within the cell. Perhaps the cells will

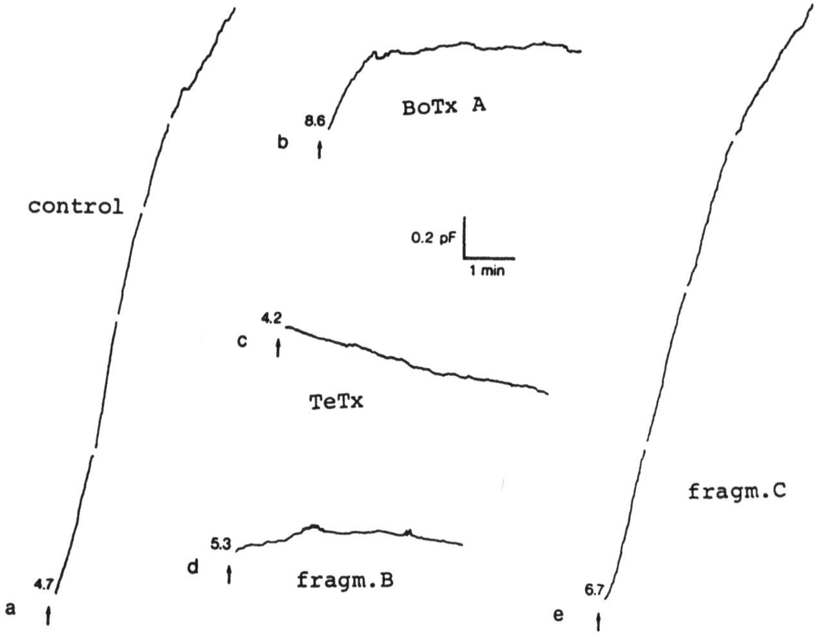

Fig. 14. Effects of clostridial neurotoxins and tetanus toxin fragments A and B on the secretory responses (measured as capacitance changes) of bovine adrenal chromaffin cells stimulated by elevation of the intracellular Ca^{2+} concentration. *a* Typical time course of capacitance changes recorded from a stimulated control cell. *b* Typical inhibitory effect of botulinum A toxin at 50 µg/ml in the pipette for 1 h. *c* Inhibition of exocytosis by tetanus toxin as in *b*. *d* Inhibition by fragment B at 25 µg/ml as in *b*. *e* Capacitance increase recorded from a cell preloaded with fragment C at 25 µg/ml in the pipette (Penner et al. 1986)

eventually present the long-sought toxin substrate too, provided the electrophysiological effect reported can be reproduced in a biochemical system. Prospects are favourable since botulinum A, B, and D toxins have been reported to inhibit catecholamine release from bovine adrenal medullary cells (Knight 1986).

Summary and Outlook

Tetanus and botulinum A toxin share many properties.

1. Their sizes and general structure, and even some stretches of the amino acids are similar.
2. They display similar pharmacokinetics with respect to binding to gangliosides and neural membranes, to internalization, and to axonal transport.
3. Although different in quantitative terms, both toxins influence many electrophysiological and neurochemical phenomena linked with neurotransmitter release on isolated organs, subcellular particles and cell cultures.

Future work is expected to reveal more details on the structure and the pharmacokinetics of the toxins. However, the search for the mode of action is just in its beginnings.

References

Aktories K, Bärmann M, Ohishi I, Tsuyama S, Jakobs K, Habermann E (1986) Botulinum C_2 toxin ADP-ribosylates actin. Nature (London) 322:390-392

Aktories K, Weller U, Chhatwal GS (1987) Clostridium botulinum type C produces a novel ADP ribosyltransferase distinct from botulinum C_2 toxin. FEBS Lett 212:109-113

Albus U, Habermann E (1983) Tetanus toxin inhibits the evoked outflow of an inhibitory (GABA) and an excitatory (D-aspartate) amino acid from particulate brain cortex. Toxicon 21:97-110

Bigalke H, Dimpfel W, Habermann E (1978) Suppression of ^3H acetylcholine release from primary nerve cell cultures by tetanus and botulinum-A toxin. Naunyn Schmiedeberg's Arch Pharmacol 303:133-138

Bigalke H, Ahnert-Hilger G, Habermann E (1981a) Tetanus toxin and botulinum A toxin inhibit acetylcholine release from but not calcium uptake into brain tissue. Naunyn Schmiedeberg's Arch Pharmacol 316:143-148

Bigalke H, Heller I, Bizzini B, Habermann E (1981b) Tetanus toxin and botulinum A toxin inhibit release and uptake of various transmitters, as studied with particulate preparations from rat brain and spinal cord. Naunyn Schmiedeberg's Arch Pharmacol 316:244-251

Bigalke H, Dreyer F, Bergey GK (1985) Botulinum A neurotoxin inhibits non-cholinergic synaptic transmission in mouse spinal cord neurons in culture. Brain Res 360:318-324

Bigalke H, Müller H, Dreyer F (1986) Botulinum A neurotoxin, unlike tetanus toxin, acts via a neuraminidase-sensitive structure. Toxicon 24:1065-1074

DasGupta BR (1981) Structure and structure-function relation of botulinum neurotoxins. In: Lewis GE (ed) Biomedical aspects of botulism. Academic Press, London New York, pp 1-19

Dolly JO, Black J, Williams RS, Melling J (1984) Acceptors for botulinum neurotoxin reside on motor nerve terminals and mediate its internalization. Nature (London) 307:457-460

Donovan JJ, Middlebrook JL (1986) Ion-conducting channels produced by botulinum toxin in planar lipid membranes. Biochemistry 25:2872-2876

Dreyer F, Schmitt A (1983) Transmitter release in tetanus and botulinum A toxin-poisoned mammalian motor endplates and its dependence on nerve stimulation and temperature. Pfluegers Arch 399:228-234

Eisel U, Jarausch W, Goretzki K, Henschen A, Engels J, Weller U, Hudel M, Habermann E, Niemann H (1986) Tetanus toxin: Primary structure, expression in *E. coli*, and homology with botulinum toxins. EMBO J 5:2495-2502

Foster JW, Kinney DM (1985) ADP-Ribosylating microbial toxins. CRC Crit Rev Microbiol 11:273-298

Habermann E (1981) Tetanus toxin and botulinum A neurotoxin inhibit and at higher concentrations enhance noradrenaline outflow from particulate brain cortex in batch. Naunyn Schmiedeberg's Arch Pharmacol 318:105-111

Habermann E, Albus U (1986) Interaction between tetanus toxin and rabbit kidney: A comparison with rat brain preparations. J Neurochem 46:1219-1226

Habermann E, Dreyer F (1986) Clostridial neurotoxins: Handling and action at the molecular and cellular level. Curr Top Microbiol Immunol 129:93-179

Habermann E, Wellhöner HH, Räker KO (1977) Metabolic fate of ^{125}I-tetanus toxin in the spinal cord of rats and cats with early local tetanus. Naunyn Schmiedeberg's Arch Pharmacol 299:187-196

Habermann E, Dreyer F, Bigalke H (1980) Tetanus toxin blocks the neuromuscular transmission in vitro like botulinum A toxin. Naunyn Schmiedeberg's Arch Pharmacol 311:33-40

Habig WH, Bigalke H, Bergey GK, Neale EA, Hardegree MC, Nelson PG (1986) Tetanus toxin in dissociated spinal cord cultures: Long-term characterization of form and action. J Neurochem 47:930-937

Hoch DH, Romero-Mira M, Ehrlich BE, Finkelstein A, DasGupta BR, Simpson LL (1985) Channels formed by botulinum, tetanus and diphtheria toxins in planar lipid bilayers: relevance to translocation of proteins across membranes. Proc Natl Acad Sci USA 82:1692-1696

Janicki P, Habermann E (1983) Tetanus and botulinum toxins inhibit, and black widow spider venom stimulates the release of methionine-enkephalin-like material in vitro. J Neurochem 41:395-402

Knight DE (1986) Botulinum toxin types A, B and D inhibit catecholamine secretion from bovine adrenal medullary cells. FEBS Lett 207:222-226

Lazarovici P, Yavin E (1986) Affinity-purified tetanus neurotoxin. Interaction with synaptic membranes: Properties of a protease-sensitive receptor component. Biochemistry 25:7047-7054

Le Vine H, Cuatrecasas P (1986) An overview of toxin-receptor interactions. In: Dorner F, Drews J (eds) Pharmacology of bacterial toxins. Pergamon, Oxford, pp 31-76

Mellanby J (1984) Comparative activities of tetanus and botulinum toxins. Neuroscience 11:29-34

Montecucco C (1986) How do tetanus and botulinum toxins bind to neuronal membranes? Trends Biochem Sci 11:314-317

Ohashi Y, Kamiya T, Fujiwara M, Narumiya S (1987) ADP-Ribosylation by type C_1 and D botulinum neurotoxins: Stimulation by guanine nucleotides and inhibition by guanidino-containing compounds. Biochem Biophys Res Commun 142:1032-1038

Penner R, Neher E, Dreyer F (1986) Intracellularly injected tetanus toxin inhibits exocytosis in bovine adrenal chromaffin cells. Nature (London) 324:76-78

Pierce EJ, Davison MD, Parton RG, Habig WH, Critchley DR (1986) Characterization of tetanus toxin binding to rat brain membranes. Evidence for a high-affinity proteinase-sensitive receptor. Biochem J 236:845-852

Roa M, Boquet P (1985) Interaction of tetanus toxin with lipid vesicles at low pH. Protection of specific polypeptides against proteolysis. J Biol Chem 260:6827-6835

Schmidt JJ, Sathyamoorthy V, DasGupta BR (1984) Partial amino acid sequence of the heavy and light chains of botulinum neurotoxin type A. Biochem Biophys Res Commun 119:900-904

Schmidt JJ, Sathyamoorthy V, DasGupta BR (1985) Partial amino acid sequences of botulinum neurotoxins Types B and E. Arch Biochem 238:544-548

Schmitt A, Dreyer F, John Ch (1981) At least three sequential steps are involved in the tetanus toxin-induced block of neuromuscular transmission. Naunyn Schmiedeberg's Arch Pharmacol 317:326-330

Shone CC, Hambleton P, Melling J (1985) Inactivation of *Clostridium botulinum* type A neurotoxin by trypsin and purification of two tryptic fragments. Proteolytic action near the COOH-terminus of the heavy subunit destroys toxin-binding activity. Eur J Biochem 151:75-82

Simpson LL (1980) Kinetic studies on the interaction between botulinum toxin type A and the cholinergic neuromuscular junction. J Pharmacol Exp Ther 212:16-21
Staub GC, Walton KM, Schnaar RL, Nichols T, Baichwal R, Sandberg K, Rogers TB (1986) Characterization of the binding and internalization of tetanus toxin in a neuroblastoma hybrid cell line. J Neurosci 6:1443-1451
Weller U, Taylor CF, Habermann E (1986) Quantitative comparison between tetanus toxin, some fragments and toxoid for binding and axonal transport in the rat. Toxicon 24:1055-1063
Wreggett KA (1986) Bacterial toxins and the role of ADP-ribosylation. J Receptor Res 95-126
Yavin E, Nathan A (1986) Tetanus toxin receptors on nerve cells contain a trypsin-sensitive component. Eur J Biochem 154:403-407

Stage-Specifically Expressed Antigens and Genes in *Trypanosoma cruzi*

G. Harth, C. G. Haidaris, and M. So[1]

Introduction

The parasitic protozoan *Trypanosoma cruzi* is the causative agent of Chagas' disease which affects some 20-30 million people in Central and South America [1]. During its life cycle, the parasite undergoes a number of morphologic and physiologic changes as it passes between two hosts, a reduviid bug and man. Epimastigote or noninfectious forms are found in the midgut of the insect. In the hindgut they change to infectious metacyclic trypomastigotes which are deposited, via the insect feces, on an individual's skin. Soon after entry through a break in the skin or via a mucous membrane, the trypomastigotes invade host cells, replicate as amastigotes, and convert back to trypomastigotes. When the cell ruptures, these forms are released in the bloodstream. Here, they either spread to surrounding tissues or enter the circulation and are taken up by the hematophagous reduviid. In the insect's gut, they convert to replicating epimastigotes and the cycle is completed [1]. The most striking change occurs as epimastigotes convert to trypomastigotes. In contrast to epimastigotes, trypomastigotes are infectious to man, do not replicate, are resistant to complement mediated lysis and express high levels of neuraminidase [1-4]. The basis for these metabolic and antigenic changes is directly linked to differential gene expression.

The signals that trigger these changes are unknown. A key to gaining an understanding of the disease is the unravelling, at the molecular level, of the processes that lead to the stage-specific expression of genes in this parasite.

Here we report on the properties of the *T. cruzi* neuraminidase, a developmentally regulated antigen of this parasite, and describe the isolation and characterization of genes specifically expressed by trypomastigotes.

Results

Purification of *T. cruzi* neuraminidase. We purified the neuraminidase (sialidase, EC 3.2.1.18) from cell membranes of metacyclic and tissue culture trypomastigotes as well as from culture supernatants. For large-scale preparations, the protein was solubilized from membranes of axenically cultured metacyclic trypomastigotes. The enzyme was purified more than 5000 fold to apparent homogeneity by several chromatography steps and size fractionation. Regardless of the source of the neuraminidase, it always migrated as a single band of M_r 60,000 in a silver stained, denaturing polyacrylamide gel (Fig. 1).

[1] Department of Molecular Biology MB-4, Research Institute of Scripps Clinic, 10666 North Torrey Pines Road, La Jolla, CA 92037, USA

38. Colloquium Mosbach 1987
Molecular Basis of Viral and Microbial Pathogenesis

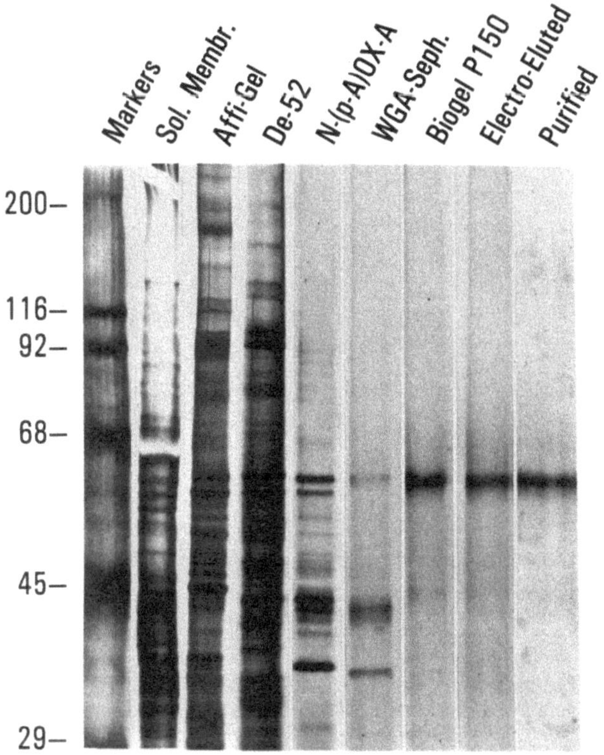

Fig. 1. Purification of *T. cruzi* neuraminidase. Active fractions were run on a denaturing 8.5% polyacrylamide gel. *Markers* molecular weight markers, in kDa, 5 µg each; *Sol. Membr.* soluble membrane extract, 5 µg; *Affi-Gel* Affi-Gel Blue, 100-200 mesh, 1 µg; *DE-52* diethylaminoethyl Cellulose DE 52, 1 µg; *N-(p-A)Ox-A* N-(p-aminophenyl)-oxamic acid-agarose, 1 µg; *WGA-Seph* wheat germ lectin-Sepharose 6MB, 1 µg; *Biogel P150* Bio-Gel P-150, 1502240, 100-200 mesh, 1 µg; *Electro-Eluted* after two rounds of electro-elution from polyacrylamide gels [6], 1 µg. Purified, after triethylamine/acetone extraction [7], 1 µg. Proteins were visualized by silver staining [18]

Analysis of Differential Expression and Activity with Various Substrates. Membrane extracts of 1 x 10^{10} epimastigotes and metacyclic trypomastigotes were prepared and chromatographed on wheat germ lectin-Sepharose 6MB. Fractions with neuraminidase activity were used in quantitative assays with 2'-(4-methylumbelliferyl)-α-D-N-acetylneuraminic acid (4MU-NANA) as the substrate. The detectable amount of enzyme increased by a factor of 17 from 11,000 µU in epimastigotes to 183,000 µU in trypomastigotes.

To determine the range of the enzyme's specificity, quantitative assays were performed with biological substrates that contained differently linked and modified sialic acid residues. For comparison, the neuraminidase of *Vibrio cholerae* [5] was included in these assays (Table 1).

These data confirmed that metacyclic trypomastigotes possess more neuraminidase activity on a per cell basis than epimastigotes. The enzyme from both forms was active on a variety of substrates. The O-acetylation at several positions in bovine submaxillary mucin obviously rendered these sialic acids less susceptible to the enzymes tested. Further addition of *T. cruzi* neuraminidase purified from metacyclic trypomastigotes did not lead to more than 50% of sialic acids being released from this substrate.

Detection of Neuraminidase with Antibodies. The purified enzyme was eluted twice from polyacrylamide gels [6,7] and used to raise antisera in rabbits. Immunoglobulins (Ig) were prepared from the antisera and their titers determined by dot blot analyses with 250 nanograms

Table 1. Release of sialic acids from various substrates by *T. cruzi* neuraminidase

Substrates	O-Substitution of sialic acids		Type of linkage of sialic acids	Type of oligosaccharide	% Sialic acids released		
					MT	EPI	VCN
4MU-NANA	---		α2'-->7NANA	---	90	6	97
α_1-Acid glycoprotein	---		α2-->3Gal	N-linked	93	6	95
Bovine submaxillary mucin	7-OAc(?),	50%	GlcNAc3 (?) ↓	O-linked	40	2.5	60
	8-OAc(?) } 9-OAc(?) }	80%	α2-->6GalNAc1				
Human red blood cells	---		α2-->3Gal or	O- and N-linked	75	5	95
			α2-->GalNAc	gangliosides			

Reaction mixtures contained 660 nmol of bound sialic acids and the enzyme units equivalent to 2.5×10^7 organisms; for MT, metacyclic trypomastigotes, 458 μU; for EPI, epimastigotes, 27 μU; for VCN, *V. cholerae* neuraminidase (Calbiochem), 500 μU. Data on sialic acid types and linkages were taken from the literature [5]; Gal, galactose; GalNAc, N-acetylgalactosamine; GlcNAc, N-acetylglucosamine; NANA, N-acetylneuraminic acid; -OAc, O-acetyl-; (?) indicates that this O-acetyl group may be present or absent on a given sialic acid residue in the protein molecule.

of partially purified membrane extracts. Antigens from metacyclic trypomastigotes were detected by antibody diluted 1:4050, antigens from epimastigotes only by an antibody dilution of 1:450. Pre-immune globulins were negative above a dilution of 1:50 (data not shown).

To analyze the specificity of the immunoglobulins in more detail, Western blots [8] were performed with partially purified membrane extracts and compared to the reaction with the purified enzyme (Fig. 2).

The protein that reacted most strongly with immunoglobulins from each of the two rabbits was an entity of M_r 60,000. The bands in the extracts correlated well with the purified molecule. The reaction with the antigens from epimastigotes revealed a very faint band at the same position. Minor cross reacting protein species could be detected in the molecular weight ranges of 90,000, 85,000, 52,000, 40,000, 36,000, 33,000 and 30,000. Ig #2963 reacted more specifically with the neuraminidase than Ig #2964. Pre-immune Ig did not recognize proteins in either extract.

Reaction of Parasite Membrane Antigens with Chagasic Sera. In the course of an infection, parasites are spread as metacyclic trypomastigotes in the feces of an infected reduviid bug [1]. These forms, like tissue culture trypomastigotes, express high levels of neuraminidase. Hence, we considered that sera from patients with Chagas' disease could contain antibodies directed against this molecule. The sera used in this study were from an acute Chagasic patient who first contracted Chagas' disease in California in August 1982. The diagnosis was verified by culturing *T. cruzi*, zymodeme 1, from the patient's blood.

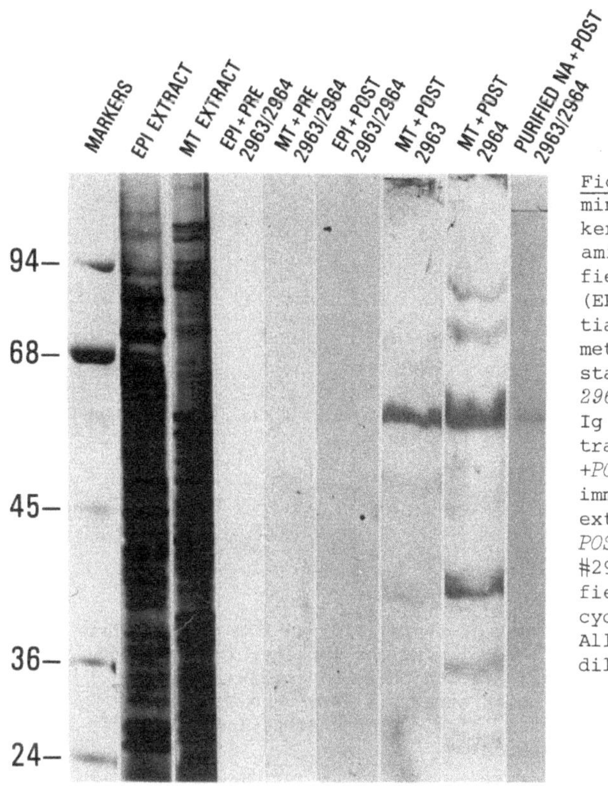

Fig. 2. Immunological detection of neuraminidase. *Markers* molecular weight markers, in kDa, 5 μg each, stained with amido black; *EPI extract* partially purified membrane extract from epimastigotes (EPI extract), 2.5 μg; *MT extract* partially purified membrane extract from metacyclics (MT extract), 2.5 μg, both stained with silver nitrate; *EPI+PRE 2963/2964* EPI extract with pre-immune Ig #2963/2964; *MT+PRE 2963/2964* MT extract with pre-immune Ig #2963/2964; *EPI +POST 2963/2964* EPI extract with post-immune Ig #2963/2964; *MT+POST 2963* MT extract with post-immune Ig #2963; *MT+POST 2964* MT extract with post-immune Ig #2964; *Purified NA+POST 2963/2964* purified neuraminidase (250 ng) from metacyclics with post-immune Ig #2963/2964. All antibodies were used at a 1:500 dilution

She responded to nifurtimox, an anti-trypanosomal agent, which lowered the anti-*T. cruzi* antibody titer from 512 to 4 over a period of 8 months [9].

We probed membrane proteins from metacyclic trypomastigotes of strain Y, zymodeme 2 [10], and strain Tulahuen, zymodeme 3 [10], with sera from this acute Chagasic patient. For comparison, partially purified membrane proteins from *Trypanosoma rangeli* d'Alessandro [11], a nonpathogenic parasite, and *Leishmania major* WR 567 [12,13], the causative agent of human cutaneous leishmaniasis, were also probed with the same sera (Fig. 3).

In general, sera taken at different times during the infection recognized the same set of antigens in strain Y. One of the predominant bands was the neuraminidase at M_r 60,000. The enzyme was in fact expressed in vivo in quantities sufficient to induce a strong antibody response. The intensity of the neuraminidase signal decreased gradually, and 8 months after infection the enzyme could no longer be detected by the individual's sera. In *T. cruzi* Tulahuen, the neuraminidase was also one of the predominant antigens. The signal intensity decreased faster than in strain Y. Eight months after initiation of chemotherapy no antigens were detected by the patient's sera.

In contrast, the antigens detected in *T. rangeli* d'Alessandro, cultured in two different media, differed from those detected in the two *T. cruzi* strains. The antibodies did not recognize any antigens in *L. major* WR 567.

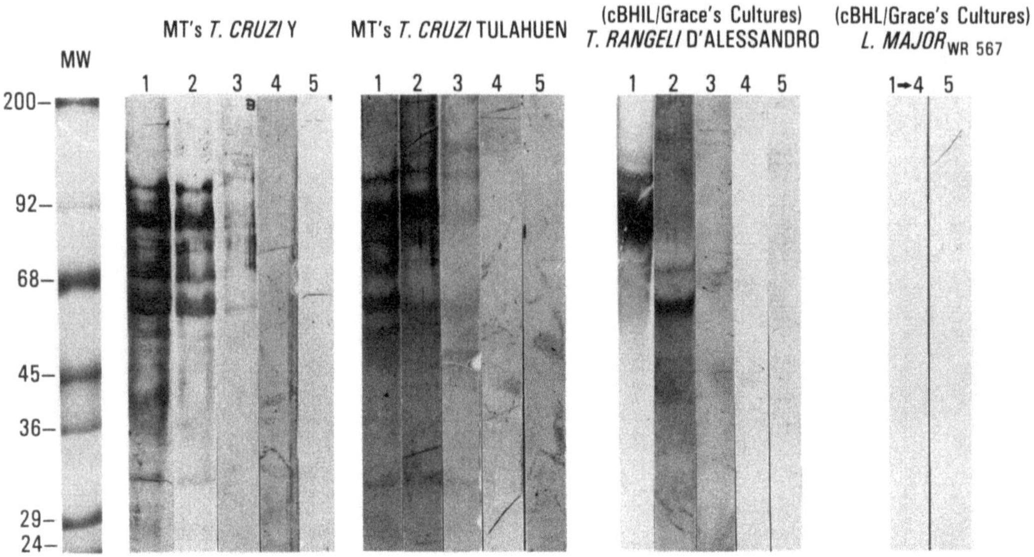

Fig. 3. Detection of partially purified membrane proteins, 2.5 µg, from various protozoan parasites with Chagasic sera [9]. *MW* molecular weight markers, in kDa, 5 µg each, stained with amido black; Chagasic sera: *1* acute phase; *2* 2 weeks later; *3* 4 weeks later; *4* 8 months later; *5* normal human sera; all sera were used at a 1:250 dilution; *MT's* metacyclic trypomastigotes; *cBHIL/Grace's* different growth media (see Experimental Procecures, cell cultures)

Cloning of Developmentally Regulated Genes from the Metacyclic Trypomastigote. We have isolated a pool of developmentally regulated genes whose expression is significantly elevated in metacyclic trypomastigotes compared to epimastigotes. Poly A^+ RNA from upper log phase epimastigotes and cultured metacyclics provided the starting material for the cloning. Single-stranded cDNA was prepared from the poly A^+ RNA of metacyclics and annealed to a 50 fold excess of poly A^+ RNA from epimastigotes. After chromatography on hydroxyapatite, only the cDNA species that did not hybridize to epimastigote poly A^+ RNA came out in the column effluent [14]. Typically, 30% of the input cDNA remained single-stranded. After synthesis of the second strand cDNA molecules were cloned into the Pst I site of pUC19 [15] using the G-C tailing method [16] and transformed into *E. coli* strain RR1 (BRL, Gaithersburg, MD).

Identification of Clones Specific for the Metacyclic Stage. From this cloning procedure, we obtained approximately 30,000 clones. Replicate nitrocellulose filters containing 5000 colonies were probed with radiolabeled poly A^+ RNA from either epimastigotes or metacyclics [14]. Clones whose signal intensity was at least ten fold greater with metacyclic mRNA compared to epimastigote mRNA were considered specific for the metacyclic forms. After four rounds of screening, 58 of approximately 5000 clones corresponded to messages whose expression was enhanced in metacyclics. Four metacyclic specific clones which bound little if any mRNA from epimastigotes were chosen for detailed analysis. Cross-hybridization studies showed that the four clones selected were unique (data not shown). We do not know if the remaining 54 clones are unique as well.

Characterization of Four Metacyclic Specific Clones. We confirmed the stage specificity of the four cDNA clones, MT-1, MT-2, MT-24, MT-49, by probing mRNA from metacyclics and epimastigotes with labeled plasmid DNA from each of the clones (Fig. 4). In each case, the hybridization signal was much stronger with mRNA from metacyclics than from epimastigotes. In addition, the relative abundance of the corresponding RNA species seemed to vary considerably among the four clones. MT-24 codes for a highly expressed message, while the RNA species corresponding to MT-2 appeared to be rare. mRNA probed with the plasmid vector or with a common clone are shown for comparison.

Cloning of the Genomic Equivalents. Our primary objective has been to identify the trypomastigote proteins for which the cDNA clones code. Since the length of the four cDNA clones was rather short, 250-400 base pairs, we isolated the corresponding full-length genes for cloning into bacterial expression vectors. A larger molecule would increase the chances that antibodies directed toward the recombinant portion of the fusion molecule would be elicited during immunization. We sequenced a portion of each of the four clones [17] and synthesized oligonucleotides (Applied Biosystems Synthesizer) for probing genomic banks of *T. cruzi* Y strain DNA. A pair of 20 base long oligonucleotides, one from each strand of cDNA, was prepared from each of the four clones. A mixture of the eight radiolabeled oligonucleotides was used to probe a library of BamH I digested genomic DNA cloned into pBS (Stratagene, San Diego, CA) and transformed into *E. coli* DH5 (BRL, Gaithersburg, MD). Screening of the genomic bank resulted in the identification of a number of genomic clones that hybridized with the oligonucleotide probes. Seven overlapping clones ranging in size from 4 to 20 kb pairs bound oligonucleotides from cDNA clone MT-1; seven overlapping clones ranging in size from 2-6 kb pairs that were specific for MT-2 were also isolated. Only one genomic clone of 0.5 kilobase pairs was obtained, corresponding to cDNA clone MT-49. No genomic counterparts for MT-24 were isolated during this particular screening (Table 2). Cross-hybridization studies of all the genomic clones confirmed once again that the cDNA clones MT-1, MT-2, and MT-49 were independent isolates which code for different gene products.

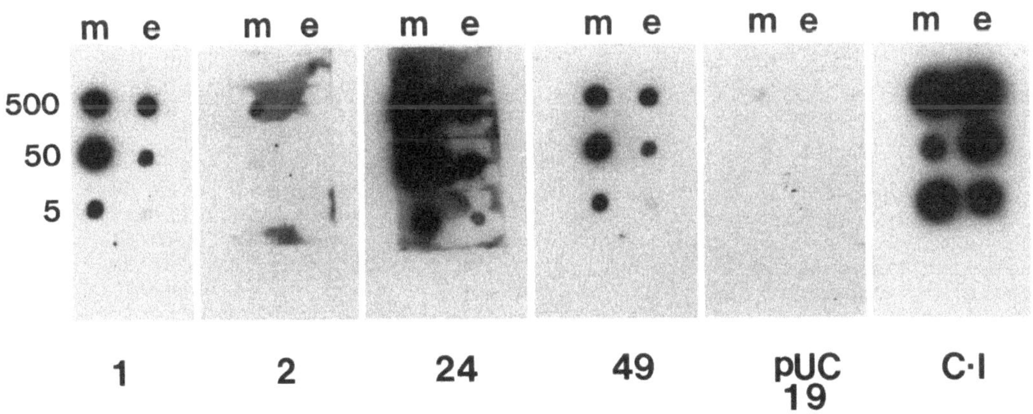

Fig. 4. Northern dot blot analysis with metacyclic specific clones [16]. mRNA, 500 ng, 50 ng, 5 ng for metacyclics (*m*) and epimastigotes (*e*) was spotted onto nitrocellulose and probed with nick translated DNA [16] from the cDNA clones MT-1 (*1*), MT-2 (*2*), MT-24 (*24*), MT-49 (*49*), 1 x 10^7 cpm/μg; *pUC19* the cloning vector, and *C-1* a cDNA clone common to metacyclics and epimastigotes were used to probe the same RNA's. Hybridization was for 16 h at 42°C in buffer containing 50% formamide. Exposure time was 72 h

Table 2. Characterization of metacyclic (MT)-specific clones

MT-specific cDNA Clone	BamH I Genomic clone	Size (kb)	Subclone(s)	Size (kb)	alpha Complementation	anti-MT Ab Detection
MT-1	4	18	Hinc II	1.4	-	-
			Hinc II	1.8	-	-
	6	10	Hinc II	0.5	+	-
	7	6	Hinc II	1.7	-	-
	18	12	Hinc II	1.7	+	+
MT-2	1	1.2	BamH I	1.2	-	-
	3	1.4	BamH I	1.4	+	-
	10	6	Hinc II	2.6	-	-
MT-49	1	0.5	BamH I	0.5	-	-

Subcloning of Genomic Fragments. Restriction enzyme digestion of the MT-1 specific genomic clones and subsequent probing of the digests with MT-1 oligonucleotides identified several small Hinc II fragments of various sizes that hybridized to the oligonucleotides. Similar studies were performed with the MT-2 specific genomic clones. From both sets of genomic clones, restriction fragments which hybridized to the oligonucleotides were eluted from polyacrylamide gels, subcloned into pBS and transformed into *E. coli* DH5alpha (BRL, Gaithersburg, MD). The inserts were cloned into the vector within the alpha fragment of the beta-galactosidase gene. This construction permits complementation with the delta M15 *lac*Z gene in DH5alpha. Subclones MT-1-6, MT-1-18, and MT-2-3 demonstrated beta-galactosidase activity, indicating the generation of a functional fusion protein. Furthermore, antibodies raised against a set of membrane proteins from metacyclics (unpublished results) detected the fusion molecule in subclone MT-1-18 (Table 2).

Discussion

As *Trypanosoma cruzi* passes through its life cycle, a number of characteristics and attributes displayed by the parasite appear to be developmentally regulated. Amastigotes and epimastigotes are replicative forms whereas the trypomastigote form does not replicate. However, only the trypomastigote can actively infect a variety of mammalian cells and acquires resistance to lysis by the alternative pathway of complement [1,2]. All the phenotypic changes described are directly linked to differential gene expression. The goal of our research has been to understand these changes at the molecular level.

The only developmentally regulated protein whose function is known is the trypomastigote specific neuraminidase. The ability to assay protein preparations for enzyme activity greatly facilitated the purification of the neuraminidase. However, the in vivo substrate of the enzyme and its role in the infectious process remain to be elucidated. The generation of an isogenic mutant in the neuraminidase gene of *T. cruzi* and the development of an efficient transformation/selection system will answer these questions in the future. We are in the process of cloning the gene for the neuraminidase using two different approaches.

The other approach taken to the study of developmentally regulated genes in *T. cruzi* has been to isolate a pool of genes that are specifically expressed by cultured metacyclic trypomastigotes. Since the parasite acquires the capacity to infect a host cell as it metamorphoses from epimastigote to trypomastigote, it has been of interest to characterize the genes whose expression is elevated in this form. The nature of the regulatory sequences that control expression of the stage specific structural genes demand analysis. We are in the process of studying these sequences in several developmentally regulated genes to determine if gene rearrangements and/or the binding of regulatory proteins to controlling DNA elements play a role in the regulation of gene expression.

In addition, it is important to identify the protein moieties encoded by these genes, since the structures could be directly involved in the infectious process. We are taking two approaches to identify the native protein coded for by the genes. We have immunized mice with the fusion protein in an effort to raise antisera against the recombinant portion of the molecule. Second, we have sequenced the clones expressing the MT-1 and MT-2 fusion molecules and are preparing peptides based on the nucleic acid sequence. Antisera to the peptides and the fusion molecules will then be used to immunologically detect the native proteins in *T. crusi*.

In conclusion, we have taken a biochemical and molecular genetic approach to the study of developmentally regulated phenotypes in the parasitic protozoan *Trypanosoma cruzi*. It is hoped that these analyses will lead to a better understanding of the pathogenic mechanisms used by the organism. This will provide the basis for the rational design of reagents that can be used for prophylaxis, diagnosis, and treatment of Chagas' disease.

Experimental Procedures

Parasite Strains. *Trypanosoma cruzi* strain Y, zymodeme 2 [10], was obtained from Dr. M. Pereira, Department of Geographic Medicine, Tufts University, Boston, MA. Dr. S. Nickell from our department provided *Leishmania major* WR 567 [13]. *Trypanosoma cruzi* Tulahuen, zymodeme 3 [10], and *Trypanosoma rangeli* d'Alessandro [11] were gifts from Dr. R. Kuhn, Department of Biology, Wake Forest University, Winston-Salem, N.C.

Cell Cultures. Epimastigotes of the two *T. cruzi* strains, *T. rangeli* d'Alessandro, and *L. major* WR 567 were grown at 27°C in Brain Heart Infusion medium, pH 7.4, supplemented as described [4].

Epimastigotes were converted to metacyclic trypomastigotes by culture for seven days at 27°C in Grace's medium [19]; 40-60% of the epimastigotes converted to metacyclics. Handling and purification of the various forms of the parasite will be described in detail elsewhere [20].

Purification of the Neuramidase. The enzyme was purified to apparent homogeneity by several chromatography steps and size fractionation. Details of the procedure will be described elsewhere [20]. Starting from 4 x 10^{10} metacyclics, 75 µg of pure neuraminidase were obtained.

Enzyme Assays. Reactions were run in 100 mM Na-acetate pH 6.8 for 20 h at 37°C. The amounts of enzyme and substrate, 2'-(4-methylumbelliferyl)-α-D-N-acetylneuraminic acid (4MU-NANA), used are given in the text. The release of NANA was followed fluorometrically [21].

Assays with other substrates, 660 nmol of bound sialic acids per reaction, were done for 20 h at 37°C as outlined [5,22].

Patient Sera. Sera from a patient, a well-documented case of Chagas' disease in California [9], were obtained from the Centers of Disease Control (CDC) in Atlanta, GA. Titers were as follows: Acute phase, 1:512; 2 weeks later, 1:256; 4 weeks later, 1:32; 8 months later, 1:4. For Western blots [8], sera were used in a 1:250 dilution.

All other procedures will be described in detail elsewhere [20,23].

Acknowledgments. This work was supported in part by the Rockeller Foundation and the NIH grant AI 22983 (M.S.), by NIH National Research Service Award AI 06837 (C.G.H.) and by grant Ha 1312/1-1 from the Deutsche Forschungsgemeinschaft (G.H.).

References

1. De Souza W (1984) Cell biology of *Trypanosoma cruzi*. Int Rev Cytol 86:197-283
2. Sher A, Hieny S, Joiner K (1986) Evasion of the alternative complement pathway by metacyclic trypomastigotes of *Trypanosoma cruzi:* dependence on the developmentally regulated synthesis of surface protein and N-linked carbohydrate. J Immunol 137:2961-2967
3. Pereira MEA (1983) A developmentally regulated neuraminidase activity in *Trypanosoma cruzi*. Science 219:1444-1446
4. Harth G, Haidaris CG, So M (1987) Purification and characterization of a neuraminidase, a developmentally regulated antigen in *Trypanosoma cruzi*. In:Agabian N, Goodman H, Nogueira N (eds) Molecular strategies of parasitic invasion; UCLA Symp Mol Cell Biol, New Ser, vol 42. Liss, New York, pp 213-222
5. Varki A, Diaz S (1983) A neuraminidase from *Streptococcus sanguis* than can release O-acetylated sialic acids. J Biol Chem 258:12465-12471
6. Hunkapiller MW, Lujan E, Ostrander F, Hood LE (1983) Isolation of microgram quantities of proteins from polyacrylamide gels for amino acid sequence analysis. Meth Enzymol 91:227-236
7. Konigsberg WH, Henderson L (1983) Removal of sodium dodecyl sulfate from proteins by ion-pair extraction. Meth Enzymol 91:254-259
8. Towbin H, Staehelin T, Gordon J (1979) Electrophoretic transfer of proteins from polyacrylamide gels to nitrocellulose sheets: procedure and some applications. Proc Natl Acad Sci USA 76:4350-4354
9. Navin TR, Roberto RR, Juranek DD, Limpakarnjanarat K, Mortenson EW, Clover JR, Yescott RE, Taclindo C, Steurer F, Allain D (1985) Human and sylvatic *Trypanosoma cruzi* infection in California. Am J Publ Health 75:366-369
10. McHardy N, Elphick JP (1978) Immunization of mice against infection with *Trypanosoma cruzi*. Cross-immunization between five strains of the parasite using freeze-thawed vaccines containing epimastigotes of up to five strains. Int J Parasitol 8:25-31
11. Tobie EJ (1961) Experimental transmission and biological comparison of strains of *Trypanosoma rangeli*. Exp Parasitol 11:1-9
12. Alexander J, Russell DG (1985) Parasite antigens, their role in protection, diagnosis and escape: The Leishmaniases. Curr Top Microbiol Immunol 120:43-67
13. Nickell SP, Gebremichael A, Hoff R, Boyer MH (1987) Isolation and functional characterization of murine T cell lines and clones specific for the protozoan parasite *Trypanosoma cruzi*. J Immunol 138:914-921
14. Davis MM, Cohen DI, Nielsen EA, Steinmetz M, Paul WE, Hood L (1984) Cell-type-specific cDNA probes and the murine I region: the localization and orientation of A_α^d. Proc Natl Acad Sci USA 81:2194-2198

15. Norrander J, Kempe T, Messing J (1983) Construction of improved M13 vectors using oligodeoxynucleotide directed mutagenesis. Gene 26:101-106
16. Maniatis T, Fritsch EF, Sambrook J (1982) Molecular cloning: a laboratory manual. Cold Spring Harbor Laboratory, Cold Spring Harbor
17. Chen E, Seeburg PH (1985) Supercoil sequencing: a fast and simple method for sequencing plasmid DNA. DNA 4:165-170
18. Wray W, Boulikas T, Wray V, Hancock R (1981) Silver staining of proteins in polyacrylamide gels. Anal Biochem 118:197-203
19. Grace TDC (1962) Establishment of four strains of cells from insect tissues grown in vitro. Nature (London) 195:788-789
20. Harth G, Haidaris CG, So M (1987) Neuraminidase from *Trypanosoma cruzi:* Analysis of enhanced expression of the enzyme in infectious forms. Proc Natl Acad Sci USA 84:8320-8324
21. Warner TG, O'Brien JS (1979) Synthesis of 2'-(4-methylumbelliferyl)-α-D-N-acetylneuraminic acid and detection of skin fibroblast neuraminidase in normal humans and in sialidosis. Biochemistry 18:2783-2787
22. Aminoff D (1961) Methods for the quantitative estimation of N-acetylneuraminic acid and their application to hydrolysates of sialomucoids. Biochem J 81:384-392
23. Haidaris CG, Harth G, So M (1987) Isolation and characterization of stage specifically expressed genes in *Trypanosoma cruzi*. (In preparation)

MIX
Papier aus verantwortungsvollen Quellen
Paper from responsible sources
FSC® C105338

If you have any concerns about our products,
you can contact us on
ProductSafety@springernature.com

In case Publisher is established outside the EU,
the EU authorized representative is:
Springer Nature Customer Service Center GmbH
Europaplatz 3, 69115 Heidelberg, Germany

Printed by Libri Plureos GmbH
in Hamburg, Germany